Biocontrol Agents and Natural Compounds against Mycotoxinogenic Fungi

Biocontrol Agents and Natural Compounds against Mycotoxinogenic Fungi

Special Issue Editors

Florence Mathieu
Selma P. Snini

MDPI • Basel • Beijing • Wuhan • Barcelona • Belgrade • Manchester • Tokyo • Cluj • Tianjin

Special Issue Editors

Florence Mathieu   Selma P. Snini
Laboratoire de Génie Chimique,   Laboratoire de Génie Chimique,
Université de Toulouse,   Université de Toulouse,
CNRS, INPT, UPS   CNRS, INPT, UPS
France   France

Editorial Office
MDPI
St. Alban-Anlage 66
4052 Basel, Switzerland

This is a reprint of articles from the Special Issue published online in the open access journal *Toxins* (ISSN 2072-6651) (available at: https://www.mdpi.com/journal/toxins/special_issues/biocontrol_natural_fungi).

For citation purposes, cite each article independently as indicated on the article page online and as indicated below:

LastName, A.A.; LastName, B.B.; LastName, C.C. Article Title. *Journal Name* **Year**, *Article Number*, Page Range.

ISBN 978-3-03936-587-6 (Hbk)
ISBN 978-3-03936-588-3 (PDF)

Cover image courtesy of Isaura Caceres, Selma P. Snini and Florence Mathieu.

Contents

About the Special Issue Editors

Florence Mathieu is a Professor at Toulouse-INP/ENSAT and she works at the Chemical Engineering Laboratory (LGC) in Toulouse. Her main research interests are microbiology, microbial interactions, and biocontrol strategies. She has been working for several years on various mycotoxins produced by several fungal genera, such as *Aspergillus* sp., i.e., aflatoxins (AFB1) and Ochratoxin A (OTA), or *Fusarium* sp. i.e., T2/HT2 toxins. Among her research themes, one concerns the reduction of the risk arising from the presence of mycotoxins in food. Her work focuses on studying mycotoxins, their biosynthetic pathways, and their elimination through processes of biological degradation/adsorption or by avoiding their production using biocontrol strategies such as the inoculation of competitive beneficial microorganisms or the use of natural compounds. Her knowledge of actinobacterial strains used as potential biocontrol agents in co-culture with filamentous mycotoxinigenic fungi will contribute to the progress of biocontrol development strategies. Florence Mathieu is an author and co-author of more than 140 international publications. She is also an expert for ANSES (the French Agency for Food, Environmental and Occupational Health & Safety) since 2018. In 2019, she was awarded the title of Officer of the French Order of Academic Palms.

Selma P. Snini received her PhD degree in Microbiology from the University of Toulouse (France, 2014) and continued as a Postdoctoral Fellow in the Toxalim Research Center in Food Toxicology (INRA, Toulouse, France) until her appointment as Assistant professor at Toulouse-INP/ENSAT in 2016. She works at the Chemical Engineering Laboratory (LGC) in Toulouse and her work focuses on the development of biocontrol strategies to reduce mycotoxin contamination. The tested strategies can include the use of natural compounds and/or microorganisms and particular attention is devoted to deciphering the mode of action of these biocontrol strategies. Thanks to her toxicology skills, Selma P. Snini is particularly interested in validating the safety of all these new biological control agents.

Editorial

Biocontrol Agents and Natural Compounds against Mycotoxinogenic Fungi

Selma Pascale Snini and Florence Mathieu *

Laboratoire de Génie Chimique, Université de Toulouse, CNRS, 31326 Toulouse, France;
selma.snini@toulouse-inp.fr
* Correspondence: florence.mathieu@toulouse-inp.fr

Received: 4 May 2020; Accepted: 26 May 2020; Published: 28 May 2020

Mycotoxins are toxic fungal secondary metabolites that contaminate food and feed. Mycotoxin contamination occurs as soon as environmental conditions are favorable for fungal growth and mycotoxin production, in the fields, during storage of raw materials and during industrial processes. To reduce mycotoxin contamination, several methods could then be adopted at these different stages. These methods can either reduce fungal growth or directly reduce the mycotoxin amount. For several years, the use of phytopharmaceutical products was favored to reduce fungal infection and thus mycotoxin contamination. However, they present numerous disadvantages such as detrimental effects in mammals, environmental contamination and subsequent strong impact on microbial biodiversity [1]. Moreover, the recurring application of fungicides could lead to the development of fungal resistance which would compromise disease control [2]. For several years, to reduce the use of such chemical products, alternative strategies based on biocontrol agents (BCAs) or natural products have been investigated.

Among mycotoxins, aflatoxin B1 (AFB1), mainly produced by *Aspergillus flavus*, is the most potent naturally occurring carcinogen and causes human hepatocarcinoma. Currently, to reduce AFB1 contamination in the fields, the use of atoxigenic strains is the most commonly used biological control method. In their study, Savić et al. isolated a native atoxigenic *A. flavus* strain from maize grown in Serbia and used it to produce a biocontrol product. The efficiency of the biocontrol product was evaluated in maize Serbian fields over two years. The results demonstrated that the biocontrol treatment had a highly significant effect in reducing total aflatoxin contamination by 73% [3]. While *A. flavus* is a saprophytic fungus, cereal crops can also be infected by phytopathogens which produce mycotoxins. Among them, the genus *Fusarium* is the most prevalent and represents a significant risk. To date, in Europe, for *Fusarium* spp, only two BCAs are available. To fill this lack of BCAs against *Fusarium* spp, Pellan et al. selected three commercial BCAs with contrasting uses and microorganism types (*Trichoderma asperellum*, *Streptomyces griseoviridis*, *Pythium oligandrum*) and studied their effect on *Fusarium graminearum* and *Fusarium verticillioides* growth and mycotoxin production. They observed variable levels of mycotoxin production and growth reduction depending on the BCA or the culture conditions, suggesting contrasting biocontrol mechanisms [4].

In addition to BCAs, microbial culture supernatants or extracts can also be used to reduce mycotoxin contamination. Indeed, microorganisms can produce several kinds of metabolites with biological activities. In this context, Zeidan et al. explored the antifungal potential of a Qatari strain of *Burkholderia cepacia* (QBC03). Their results demonstrate that this strain exhibits antifungal activity against a wide range of fungi belonging to the *Aspergillus*, *Fusarium* and *Penicillium* genera. Moreover, the addition of the *B. cepacia* culture supernatant (2.5% to 15.5%) in the culture medium drastically reduces the fungal growth of *Penicillium verrucosum*, *Aspergillus carbonarius* and *Fusarium culmorum*. Further studies will be conducted to decipher the precise mechanism of action of the antifungal compounds secreted by this *B. cepacia* strain [5]. In the same way, Hanif et al. demonstrated that fengycin extracted from *Bacillus amyloliquefaciens* FZB42 inhibits *F. graminearum* growth and mycotoxin production [6]. Similar to microbial metabolites, natural compounds can also affect fungal growth and mycotoxin production. They are extracted from plants and they can be used as aqueous extracts, organic extracts or essential oils. Wang et al. demonstrated that citral essential oil completely suppressed the mycelial growth of *Alternaria alternata* at the concentration of 222.5 µg/mL, which is the minimal inhibitory concentration (MIC). Moreover, the 1/2 MIC of this essential oil inhibits more than 97% of the mycotoxin amount. A comparative transcriptomic analysis of *A. alternata* treated or untreated revealed that citral affects transcription of genes involved in alternariol biosynthesis [7]. In the same way, Degola et al. investigated the biological activity of *Citrullus colocynthis* stem, leaf and root extracts on *A. flavus*. Among the tested tissues, leaf and root extracts showed the highest levels of AFB1 reduction (up to 80% reduction) [8].

BCAs can also be applied during industrial processes to limit fungal growth and mycotoxin contamination. As an example, in the brewing process, *Geotricum candidum*, a filamentous yeast is used to reduce *Fusarium* spp. growth and the T-2 toxin concentration. Kawtharani et al. demonstrated that *G. candidum* produces phenyllactic acid at the early stages of growth, which is responsible for the reduction of the T-2 toxin concentration through the reduction in *Fusarium* spp. growth [9].

The last scientific article included in this Special Issue is on the fringe of the other articles and deals with the use of fullerol nanoparticles (FNP) to modulate the secondary metabolite profile of the most relevant foodborne mycotoxigenic fungi belonging to the genera *Aspergillus*, *Fusarium*, *Alternaria* and *Penicillium*. This is a preliminary study to present the proof of concept for the use of FNP against mycotoxin contamination. Thus, Kovac et al. demonstrated that exposure to FNP leads to the reduction in concentrations of 35 secondary metabolites depending on the concentration of the applied FNP and the fungal genus [10].

Acknowledgments: We express our gratitude to all contributing authors and reviewers.

Conflicts of Interest: The authors declare no conflict of interest.

References

1. Zubrod, J.; Bundschuh, M.; Arts, G.; Brühl, C.; Imfeld, G.; Knäbel, A.; Payraudeau, S.; Rasmussen, J.; Rohr, J.; Scharmüller, A.; et al. Fungicides—an Overlooked Pesticide Class? *Environ. Sci. Technol.* **2019**, *53*, 3347–3365. [CrossRef] [PubMed]

2. Lucas, J.A.; Hawkins, N.J.; Fraaije, B.A. The Evolution of Fungicide Resistance. In *Advances in Applied Microbiology*; Elsevier Ltd: Amsterdam, The Netherlands, 2015; Volume 90, pp. 29–92.

3. Savic, Z.; Jaji, I.; Stankov, A.; Vukoti, J. Biological Control of Aflatoxin in Maize Grown in Serbia. *Toxins* **2020**, *12*, 162. [CrossRef] [PubMed]

4. Pellan, L.; Durand, N.; Martinez, V.; Fontana, A.; Schorr-Galindo, S.; Strub, C. Commercial biocontrol agents reveal contrasting comportments against two mycotoxigenic fungi in cereals: *Fusarium graminearum* and *Fusarium verticillioides*. *Toxins* **2020**, *12*, 152. [CrossRef] [PubMed]

5. Zeidan, R.; Ul-Hassan, Z.; Al-Thani, R.; Migheli, Q.; Jaoua, S. In-vitro Application of a Qatari *Burkholderia cepacia* strain (QBC03) in the Biocontrol of Mycotoxigenic Fungi and in the Reduction of Ochratoxin A biosynthesis by *Aspergillus carbonarius*. *Toxins* **2019**, *11*, 700. [CrossRef] [PubMed]

6. Hanif, A.; Zhang, F.; Li, P.; Li, C.; Xu, Y.; Zubair, M.; Zhang, M.; Jia, D.; Zhao, X.; Liang, J.; et al. Fengycin produced by *Bacillus amyloliquefaciens* FZB42 inhibits *Fusarium graminearum* growth and mycotoxins biosynthesis. *Toxins* **2019**, *11*, 295. [CrossRef] [PubMed]

7. Wang, L.; Jiang, N.; Wang, D.; Wang, M. Effects of essential oil citral on the growth, mycotoxin biosynthesis and transcriptomic profile of alternaria alternata. *Toxins* **2019**, *11*, 553. [CrossRef] [PubMed]

8. Degola, F.; Marzouk, B.; Gori, A.; Brunetti, C.; Dramis, L.; Gelati, S.; Buschini, A.; Restivo, F.M. *Aspergillus flavus* as a model system to test the biological activity of botanicals: An example on *citrullus colocynthis* L. schrad. organic extracts. *Toxins* **2019**, *11*, 286. [CrossRef] [PubMed]

9.   Kawtharani, H.; Snini, S.P.; Heang, S.; Bouajila, J.; Taillandier, P.; Mathieu, F.; Beaufort, S. Phenyllactic acid produced by *Geotrichum candidum* reduces *Fusarium sporotrichioides* and *F. langsethiae* growth and T-2 toxin concentration. *Toxins* **2020**, *12*, 209. [CrossRef] [PubMed]

10.   Kovac, T.; Šarkanj, B.; Borišev, I.; Djordjevic, A.; Jovic, D.; Loncaric, A.; Babic, J.; Jozinovi, A.; Krska, T.; Gangl, J.; et al. Fullerol $C_{60}(OH)_{24}$ Nanoparticles Affect Secondary Metabolite Profile of Important Foodborne Mycotoxigenic Fungi *In Vitro*. *Toxins* **2020**, *12*, 213. [CrossRef] [PubMed]

Article

Biological Control of Aflatoxin in Maize Grown in Serbia

Zagorka Savić [1], Tatjana Dudaš [1,*], Marta Loc [1], Mila Grahovac [1], Dragana Budakov [1], Igor Jajić [1], Saša Krstović [1], Tijana Barošević [1], Rudolf Krska [2,3], Michael Sulyok [2], Vera Stojšin [1], Mladen Petreš [1], Aleksandra Stankov [1], Jelena Vukotić [1] and Ferenc Bagi [1]

[1] Faculty of Agriculture, University of Novi Sad, 21000 Novi Sad, Serbia; zagorka.savic@polj.uns.ac.rs (Z.S.); marta.loc@polj.uns.ac.rs (M.L.); mila@polj.uns.ac.rs (M.G.); dbudakov@polj.uns.ac.rs (D.B.); igor.jajic@stocarstvo.edu.rs (I.J.); sasa.krstovic@stocarstvo.edu.rs (S.K.); tijana.doroski@gmail.com (T.B.); stojsinv@polj.uns.ac.rs (V.S.); mladen.petres@polj.uns.ac.rs (M.P.); aleksandra.stankov@polj.uns.ac.rs (A.S.); jelena.medic@polj.uns.ac.rs (J.V.); bagifer@polj.uns.ac.rs (F.B.)

[2] Institute of Bioanalytics and Agro-Metabolomics, Department IFA-Tulin, University of Natural Resources and Life Sciences Vienna (BOKU), A-3430 Tulln, Austria; rudolf.krska@boku.ac.at (R.K.); michael.sulyok@boku.ac.at (M.S.)

[3] Institute for Global Food Security, School of Biological Sciences, Queens University Belfast, University Road, Belfast BT7 1NN, UK

* Correspondence: tatjana.dudas@polj.uns.ac.rs

Received: 20 January 2020; Accepted: 3 March 2020; Published: 5 March 2020

Abstract: *Aspergillus flavus* is the main producer of aflatoxin B1, one of the most toxic contaminants of food and feed. With global warming, climate conditions have become favourable for aflatoxin contamination of agricultural products in several European countries, including Serbia. The infection of maize with *A. flavus*, and aflatoxin synthesis can be controlled and reduced by application of a biocontrol product based on non-toxigenic strains of *A. flavus*. Biological control relies on competition between atoxigenic and toxigenic strains. This is the most commonly used biological control mechanism of aflatoxin contamination in maize in countries where aflatoxins pose a significant threat. Mytoolbox Af01, a native atoxigenic *A. flavus* strain, was obtained from maize grown in Serbia and used to produce a biocontrol product that was applied in irrigated and non-irrigated Serbian fields during 2016 and 2017. The application of this biocontrol product reduced aflatoxin levels in maize kernels (51–83%). The biocontrol treatment had a highly significant effect of reducing total aflatoxin contamination by 73%. This study showed that aflatoxin contamination control in Serbian maize can be achieved through biological control methods using atoxigenic *A. flavus* strains.

Keywords: aflatoxin; *Aspergillus flavus*; biological control; atoxigenic strain; maize; Serbia

Key Contribution: Recently, changing climate conditions have become favourable for aflatoxin contamination of maize in Serbia and other European countries. Therefore, it is necessary to improve the available methods for managing aflatoxin contamination. This article describes research involving biological control of aflatoxin in Serbian maize using native atoxigenic isolates. Selected native atoxigenic isolate efficiently reduced aflatoxin contamination in maize.

---

1. Introduction

Aflatoxins are the most common contaminants of important agricultural commodities including maize, cottonseed, peanuts, and pistachio nuts. *Aspergillus flavus* and related species produce aflatoxins, which are secondary metabolites that can adversely affect human health and food security in warm agricultural areas [1,2]. Aflatoxins are potent, naturally-occurring carcinogens that can suppress the immune system and induce hepatocellular carcinoma, which can cause mortality in humans and livestock [3,4]. Aflatoxin B1 (AFB1), which is classified as a Group 1a carcinogen by the International Agency for Research on Cancer [5], is the most common and toxic of the four major aflatoxins B1, B2, G1, and G2. Aflatoxin concentration in food and feed is strictly regulated by international organisations due to the significant impacts on human health [6]. Products contaminated by aflatoxins have limited value and access to markets, resulting in significant economic losses [3].

Environmental and biological factors, such as increased temperatures, droughts, pest damages, host susceptibility to infection, and the aflatoxin-producing potential of fungi, have combined to raise aflatoxin contamination levels above regulated limits [7,8]. Aflatoxin contamination can occur after crop maturity when the crops are exposed to high temperature and humidity levels, which are conducive to fungal infection and may result in an increase of aflatoxin accumulation [9]. Therefore, aflatoxin contamination can start and continue after cropping, but also during storage, transport, processing, and handling [10].

In Serbia during the summer of 2012 the appearance of *A. flavus* in maize was a result of extremely stressful environmental conditions that included high air temperatures and little precipitation, and was reported as an emerging food and feed safety threat. Natural occurrences of aflatoxins are uncommon under Serbia's typical climatic conditions; however, because mycotoxin occurrence is climate-dependent [11], recent climate changes have become significant causal agents of food and feeds safety issues in Serbia [12,13].

Aflatoxins have received considerable research due to global consumer concerns related to the presence of aflatoxins in the food supply. Understanding the biology, epidemiology, and occurrence of aflatoxin-producing fungi and the development of advanced technologies for reducing these fungi are urgently needed. Biological control methods based on the competitive exclusion of toxigenic strains by atoxigenic strains have been developed as an innovative strategy for reducing aflatoxin accumulation. Atoxigenic strains displace aflatoxin producers by competing with other toxigenic and atoxigenic strains for infection sites and essential nutrients during crop development. This competition has resulted in significant reductions in aflatoxin contamination in harvested grains [14–17].

Researching the genetic diversity within aflatoxin-producing fungi population in Serbia is highly important for determining the etiology of aflatoxin contamination and for optimising the selection of native atoxigenic *A. flavus* genotypes that can be used for aflatoxin biological control products targeted to local agroecosystems. The diverse environmental conditions and soil microbiomes that are found in different locations will favour the use of native *A. flavus* strains for local agroecosystems. Thus, native *A. flavus* strains are expected to provide better results than introduced exotic genotypes [18]. This paper highlights recent research in Serbia that was designed to improve the management of aflatoxin contamination of maize using native atoxigenic isolates. Isolates of selected native atoxigenic genotypes were applied to maize crops prior to flowering as a biological control method with the goal of reducing aflatoxin contamination.

2. Results

2.1. Monitoring Deletions in the Aflatoxin Biosynthesis Isolate Mytoolbox Af01 Gene Cluster

Twenty fungal isolates determined to be *A. flavus* were subjected to Cluster Amplification Patterns (CAP) analysis for screening of missing regions in the aflatoxin biosynthesis gene cluster [19]. Four multiplex PCRs were designed to amplify 32 markers spaced at 5 kb intervals. The results of the CAP multiplex analysis revealed two atoxigenic strains, Mytoolbox Af01 (from Pivnice, Serbia) and T7/I11, whereas the other isolates were toxigenic (Figure 1). Only one isolate (Mytoolbox Af01) was chosen for biocontrol agent preparation because the two detected atoxigenic strains originated from a narrow geographical region and were genetically identical. The amplification products of the first two multiplex reactions are shown on the gel images, aligned to a schematic diagram of chromosome 3 containing the aflatoxin biosynthesis gene cluster. The CAP analyses showed that the missing cluster in the Mytoolbox Af01 strain and the second atoxigenic strain were over 40 kb. The deletion spans markers AC04 to AC11, which corresponds to the region between genes aflT and verA/aflN (Figure 2).

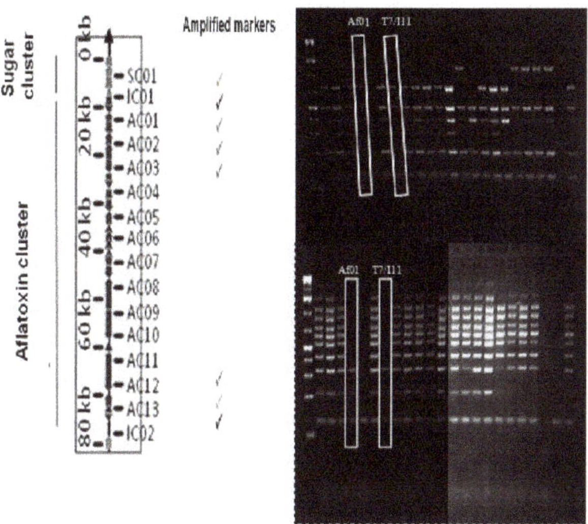

Figure 1. Images of multiplex PCR products aligned to a schematic diagram from Callicot and Cotty [19] of chromosome 3 containing the aflatoxin cluster.

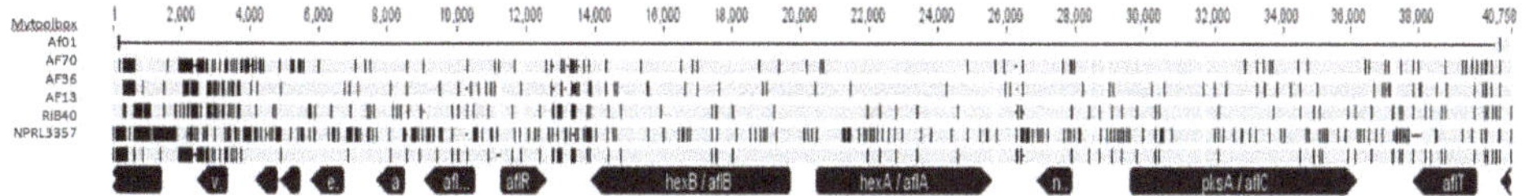

Figure 2. Comparative view of the missing 40 kb region in atoxigenic strain (Mytoolbox Af01), toxigenic *A. flavus* strains (AF70, AF13, NPRL3357), atoxigenic *A. flavus* strain (AF36), and atoxigenic *A. oryzae* strain (RIB40).

2.2. Quality Control of Atoxigenic Product

Visual evaluations of sporulation observed abundant sporulation on all tested seeds.

2.3. Intensity of A. flavus Infection in Maize

Aspergillus ear rot infections in the fields were low each year. Statistical analyses showed that there were no significant differences among treatments (at a 95% confidence level).

2.4. AFB1 Content

Significant differences in AFB1 content were found between treated and control plots in 2016 and 2017. All 2016 samples from the control plots with and without irrigation contained aflatoxin. A highly significant effect of biocontrol treatment was observed during 2016 and 2017, with an overall reduction of 73%. The reductions achieved in 2016 and 2017 were 83% and 51%, respectively.

In Sombor, the total mycotoxin contamination had mean AFB1 contents of 7.19 ppb and 3.80 ppb in 2016 and 2017, respectively, with no significant differences being found. Similarly, correlation analyses of AFB1 content in 2016 and 2017 were not significant.

We assumed irrigation would help plants avoid infection with the toxigenic *A. flavus*, which should have resulted in lower contamination levels. However, this assumption was incorrect because mean AFB1 content was higher, but not significantly different, in the Sombor treated irrigated plots in 2016 and 2017. However, the AFB1 contents in the unirrigated control plots were slightly, but not significantly, higher. The AFB1 data were logarithmically transformation and analysed using a mixed ANOVA, with 2016 and 2017 considered as repeated measurements. This method enabled analysis of biocontrol and irrigation as main effects.

Table 1 presents t-test results for equality of means and shows that application of the biocontrol agent significantly affected AFB1 levels. Irrigation had no significant effects.

Table 1. T-test for equality of means of different AFB1 contamination levels.

Effect	N	Mean	Standard Deviation	*t*-Test for Equality of Means		
				t	df	Significance Probability (2-Tailed)
Biocontrol						
Treated	32	2.31	6.713	−3.858	62	0.000
Untreated	32	8.68	6.483			
Irrigation						
Irrigated	32	5.75	8.215	0.279	62	0.782
Unirrigated	32	5.24	6.355			

Results of the multivariate analyses are shown in Figures 3 and 4. The tests showed that changes in mycotoxin AFB1 contamination with time only had significant interactions with the biocontrol treatment ($p = 0.04$, Wilks' Lambda = 0.746, df = 1, F = 9.554, $\eta_p^2 = 0.254$). Contamination levels were also significantly affected by year ($p = 0.013$, Wilks' Lambda = 0.799, df = 1, F = 7.031, $\eta_p^2 = 0.201$). The interactions of year and irrigation ($p = 0.679$, Wilks' Lambda = 0.994, df = 1, F = 0.175, $\eta_p^2 = 0.006$), plus year, biocontrol, and irrigation, were not significantly affected by contamination ($p = 0.779$, Wilks' Lambda = 0.997, df = 1, F = 0.80, $\eta_p^2 = 0.003$). Figure 3 shows AFB1 contamination levels in the control plots being heavily depended on the weather conditions during 2016 and 2017, whereas contamination levels in the treated plots remained low and stable in both years.

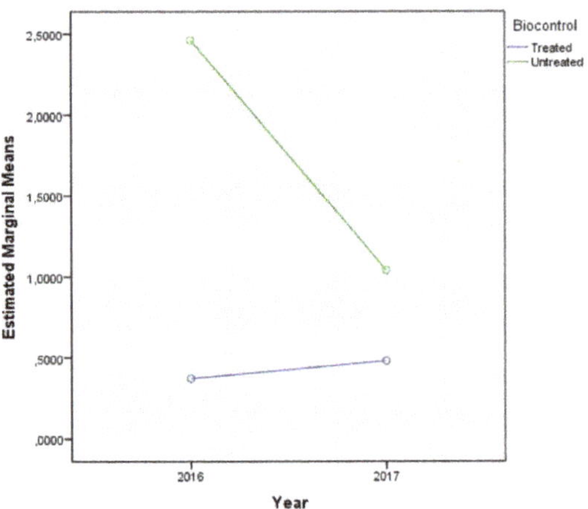

Figure 3. Multivariate test showing the influence of biocontrol on AFB1 contamination levels in 2016 and 2017.

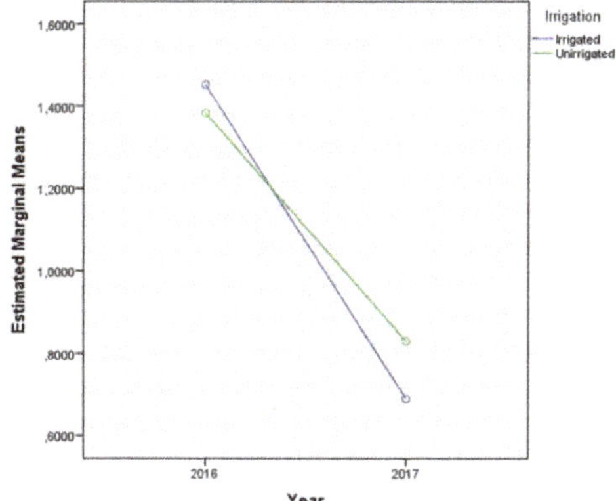

Figure 4. Multivariate test showing the influence of irrigation on AFB1 contamination levels in 2016 and 2017.

A between-subject ANOVA (Table 2) for both years showed significant effects in biocontrol application, whereas effects were not significant for irrigation, or biocontrol combined with irrigation.

Table 2. Between-subject ANOVA test of the influence of different factors on AFB1 contamination levels.

Effect	Type III Sum of Squares	df	Mean Square	F	Significance Probability	Partial Eta Squared
Intercept	75.913	1	75.913	98.064	0.000	0.778
Biocontrol	28.057	1	28.057	36.244	0.000	0.564
Irrigation	0.020	1	0.020	0.026	0.873	0.001
Biocontrol*Irrigation	0.342	1	0.342	0.442	0.512	0.016
Error	21.675	28	0.774			

2.5. Climate Conditions

In Sombor during 2016 (Figure 5), precipitation levels in the months during the vegetation period, except for April, were above the multiannual monthly average. Average daily air temperatures were above the long-term average during June and July.

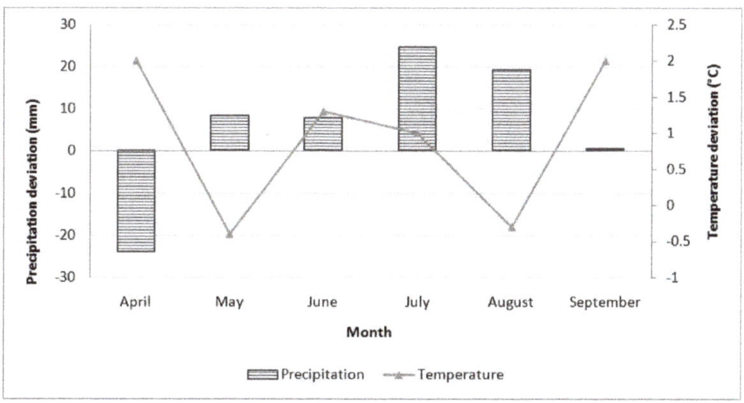

Figure 5. Deviation of total precipitation (columns) and average daily air temperature (lines) from the multiannual average (1981–2010) in Sombor during 2016.

Total rainfall during 2017 in Sombor (Figure 6) was lower than the multiannual average in April, June, and August; however, July rainfall was higher than the average. Average daily air temperatures was the long-term average from May till August.

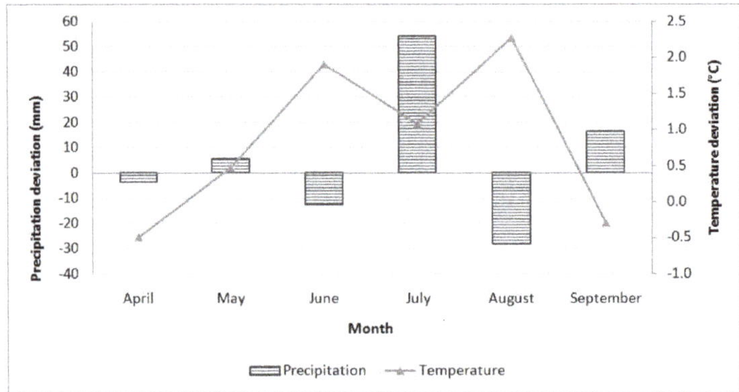

Figure 6. Deviation of total precipitation (columns) and average daily air temperature (lines) from the multiannual average (1981–2010) in Sombor during 2017.

3. Discussion

Mytoolbox Af01 is the first native atoxigenic *A. flavus* strain used as a biocontrol agent to mitigate aflatoxin contamination in Serbian maize fields. The atoxigenic potential of this strain was confirmed by the CAP analyses, which revealed the deletion of 40 kb from the aflatoxin biosynthesis gene cluster. Different types of large deletions in the aflatoxin biosynthesis gene cluster occur often [19–22], but degeneration of this gene cluster can also be caused by multiple smaller deletions or SNP mutations [23].

The biological control product based on the Mytoolbox Af01 strain was applied in one Serbian location during two consecutive years to assess its aflatoxin reduction potential. *Aspergillus* ear rot was examined every year, but the natural infection rates were low and no significant differences were observed among treatments. AFB1 content in collected samples was determined using a DAS (double antibody sandwich) ELISA test, which was found to be precise in terms of reproducibility. AFB1 was detected in samples from 2016 and 2017, even though there were no visible symptoms of *Aspergillus* ear rot on maize cobs during harvest. Sometimes, seemingly healthy maize grains can contain small levels of AFB1 [24,25]. These results point to the differences between symptomatic and asymptomatic plants and indicates that AFB1 content needs to be investigated further in environmental conditions specific to Serbian maize growing areas. Aflatoxin contamination levels were significantly affected by the application of the bioproduct. Application of the biocontrol product led to an overall mean reduction of 73% in AFB1 levels (83% in 2016 and 51% in 2017). The reduction of aflatoxin contamination from similar biocontrol products has been reported previously. In the USA, Dorner et al. [26] successfully applied an *A. flavus* atoxigenic strain that reduced aflatoxin contamination up to 87%. Moreover, four native atoxigenic strains applied in Nigeria reduced aflatoxin content 67–95% in treated crops [27]. Abbas et al. [28] reported 65–94% reductions of aflatoxin in maize ears inoculated with atoxigenic and toxigenic isolates.

Irrigation in the current experiment did not influence aflatoxin contamination in either year, in contrast to other studies that found irrigation reducing aflatoxin contamination [29–32]. These contradicting results could have been caused by different methodologies being used, timing and amount of irrigation, plant genotype, weather conditions, soil type, deficiency of easily accessible water in the soil, or other factors. In the control plots, AFB1 contamination levels were dependent on climate conditions during the different years, whereas contamination levels in the treated plots low and stable in both years.

4. Conclusions

Mytoolbox Af01 is the first native atoxigenic strain used as a biological preparation to produce significant reductions in aflatoxin levels in a Serbian field. The results indicate that the biocontrol product has a high potential for reducing aflatoxin contamination in local environmental conditions.

5. Materials and Methods

5.1. Selection of Aspergillus flavus Strains

A selection of *A. flavus* isolates (from maize in Serbia) that were characterised based on colony and spore morphology were selected for this study. Isolates were grown on 5–2 medium [33], made of V-8^{TM} juice (vegetable juice from eight vegetables) containing 2% NaCl and grown for 8–10 days at 31 °C. PCR analyses using species-specific primer Aflafor and universal reverse primer Bt2b [34] were performed to confirm identification. Reactions consisted of: 2 µL of DreamTaq Buffer, 4 µL of dNTP mix, 2 µL of each primer, 5 µL of DNA-free water, 0.2 µL of DreamTaq DNA polymerase and 1 µL DNA template. The samples were subjected to 3 min at 94 °C; 35 cycles of 30 s 94 °C, 30 s 64 °C, 20 s 72 °C; followed by 2 min at 72 °C. The products were visualised on 1% agarose gel in 0.5 x TAE buffer.

5.2. Monitoring Deletions in the Aflatoxin Biosynthesis Isolate Mytoolbox Af01 Gene Cluster

Cluster Amplification Patterns (CAP) analyses were performed for screening missing regions in the aflatoxin biosynthesis gene cluster, according to the method by Callicot and Cotty [19]. Four multiplex PCRs were designed to amplify 32 markers. Each 10 µl pre-amplification reaction contained: 0.08 µmol^{-1} of each primer, 1 × AccuStart II PCR SuperMix (Quanta Biosciences, Gaithersburg, MD, USA) and 6 ng genomic DNA. PCR reactions were carried out with the following thermal profile: 94 °C for 1 min, followed by 30 cycles of 94 °C for 30 s, 62 °C for 90 s, 72 °C for 90 s and the final extension step of 72 °C for 10 min. Products were visualised on 1.4% agarose in 1× sodium boric acid buffer [35].

5.3. Biocontrol Product Preparation

A biocontrol product with atoxigenic *A. flavus* strain was produced according to the modified method described by Garber et al. [36]. Baked sorghum seeds were used as the inoculum carrier. Prior to inoculation, atoxigenic *A. flavus* strain was cultivated on 5–2 media and incubated at 31 °C for five days. Spore production was performed on sorghum seeds with moisture levels adjusted to 20% using spores from five-day-old cultures in a suspension that was added to autoclaved and cooled sorghum seeds. Flasks were sealed with sterile Tyvek membrane to control humidity levels but allow gas exchange, and incubated for seven days at 35 °C. The spore suspension for biocontrol production was prepared by harvesting spores with 100 mL of sterile 0.5% Tween-80 solution, and a concentration of conidia adjusted to 1–5 × 10^8 per ml using a haemocytometer. The final suspension was mixed with the sorghum seeds, a seed polymer, and a dye. The moisture content of the final product was adjusted to 10%. The dye was used to indicate the sorghum seeds treated with the atoxigenic *A. flavus*.

5.4. Quality Control of the Atoxigenic Biocontrol Product Following Cotty (pers. comm.)

Prior to application, the quality of the biocontrol product was measured by development and sporulation of Mytoolbox Af01 on sorghum seeds. Individual sorghum seeds of the final biocontrol product were placed in a multi-well plate that had sterile water poured in the outer wells to increase humidity and promote sporulation. Plates were incubated in a closed plastic container at 31 °C for 7 days. Sporulation was visually recorded [37].

5.5. Sowing Maize and Application of the Atoxigenic Isolate

The biocontrol product based on atoxigenic *A. flavus* was applied to one maize hybrid (Kerbanis FAO class 500). This hybrid belongs to the FAO group of maize that farmers sow on large areas in

Serbia. The maize seeds were planted over two years in one location in Sombor (Serbia). Sowing was completed on 24th April and 11th April in 2016 and 2017, respectively.

When the plants within the inoculated plots had developed 10 leaves, the atoxigenic isolate was manually applied at 10 kg/ha to the soil surface next to each plant. In both years the biocontrol product was applied in the fields with and without irrigation. Irrigation was sufficient to keep the soil at an optimal moisture level for normal plant growth before and after bioproduct application. A drip irrigation system was used between plant rows. The total water volume applied during the vegetation period was 95 l/m^2 in 2016 and 140 l/m^2 in 2017. Each of the four combinations irrigated with bioproduct application, irrigated control (without bioproduct application), unirrigated with bioproduct application, and unirrigated control (without bioproduct application) were repeated eight times on individual plots that were over 50 m^2.

5.6. Evaluation of Intensity of A. flavus Infection in Maize

The intensity of *A. flavus* infection on maize cobs was visually evaluated 7–10 days before harvest each year. Disease intensity was evaluated by rating 100 randomly chosen ears within each plot (32 individual plots) using a scale from 1 to 7 [38]. Each ear was evaluated based on the percentage of infected kernels: 1) ear without symptoms, 2) 1–3% infected kernels, 3) 4–10% infected kernels, 4) 11–25% infected kernels, 5) 26–50% infected kernels, 6) 51–75% of infected kernels, 7) 76–100% infected kernels.

5.7. Harvest and Samples Preparation

The maize were harvested each year in September. After the harvest, 2 kg samples were taken from 32 individual plots. About 100 g of laboratory samples were prepared by grinding in a laboratory mill with pore diameter of 0.8 mm until >93% of the sample passed through the sieve. The sample was next homogenised by mixing and packed into plastic bags. Samples were stored in a freezer at −20 °C until analysis. Prior to each analysis, the samples were allowed to reach room temperature. Afterwards, the AFB1 content was determined by the ELISA test.

5.8. ELISA Test

Exactly 20 g of each ground sample was weighed in a 150 mL beaker. Aflatoxin B1 was extracted with 100 mL of 70% methanol solution on an Ultra Turrax T18 homogeniser (IKA, Staufen, Germany) for 3 min at 11,000 rpm. The crude extract was filtered through a quantitative slow filtration filter paper (Filtros Anoia, Barcelona, Spain).

The immunochemical analysis was performed using the AgraQuant®Aflatoxin B1, Quantitative Test Kit (Romer Labs, Tulln, Austria) with four calibration standard solutions (0, 2, 5, 20, and 50 ppb). The analytical procedure was carried out according to the manufacturer's instructions. Aflatoxin quantification was done on an ELISA reader equipped with a 450 nm filter (BioTec Instruments, USA). To ensure the quality of the results, the aflatoxin was validated in the laboratory. Validation parameters were evaluated according to the European Commission [39]. The limit of quantification (LOQ) of 2 µg/kg that was established by the manufacturer, was experimentally verified by analysing blank samples of corn that were treated with 2 µg/kg of an aflatoxin B1 standard solution (Sigma Aldrich, St. Louis, MO, USA). Samples containing less than 2 µg/kg aflatoxin were considered negative. The average trueness of this method was calculated by analysing eight successive corn samples from a certified reference material (CRM) coded TR-A100 (Trilogy lab, Washington, MO, USA). The average trueness of 105.4% was within acceptable limits according to the European Regulation [39]. In-house reproducibility was assessed after eight successive CRM tests, on two different occasions, resulting in 16 tests. The obtained HORRAT$_R$ value of 0.20 was within the acceptable criteria for reproducibility established by the European Regulation [39].

5.9. Statistical Analyses

Statistical analyses of the scoring data from the field maize infection and AFB1 content were performed in Statistica v. 13 (TIBCO Software Inc., Silicon Valley, CA, USA, 2017). Because the maize infection data were non-parametric, Kruskal–Wallis tests were used to determine if the mean ranks of the infection levels were the same in all treatments. Furthermore, multiple comparisons of mean ranks were used as post hoc tests to determine the treatments that were significantly different from each other. The aflatoxin content data were logarithm transformation and used in an analysis of variance to determine differences between treatments, and a multivariate test to assess between-subject effects.

5.10. Climate Conditions

Aflatoxin contamination in maize is correlated with plant, drought, and heat stress [40]. Therefore temperature and precipitation were monitored during the vegetation growth period at the locality every year. Data were obtained from a Metos®automatic weather station (Metos®, Pessl Instruments, Weiz, Austria) and compared to multiannual averages for 1981–2010 [41].

Author Contributions: Conceptualization, Z.S., M.G., D.B. and F.B.; Data curation, Z.S., T.D., M.L., T.B., M.P., A.S. and J.V.; Formal analysis, Z.S., T.D., M.L., S.K. and M.S.; Funding acquisition, I.J., R.K. and F.B.; Investigation, Z.S., T.D., M.L., T.B., M.P., A.S. and J.V.; Methodology, M.G., D.B., I.J., R.K. and M.S.; Project administration, I.J., R.K., V.S. and F.B.; Resources, I.J., R.K., V.S. and F.B.; Software, M.G. and D.B.; Supervision, M.G., D.B., I.J., R.K., V.S. and F.B.; Validation, I.J., S.K., R.K. and M.S.; Visualization, T.B., M.P., A.S. and J.V.; Writing-original draft, Z.S., T.D., M.L. and S.K.; Writing-review & editing, M.G., D.B., I.J., V.S. and F.B. All authors have read and agreed to the published version of the manuscript.

Funding: This research was funded by MyToolBox (EU's Horizon 2020 research and innovation programme, agreement No 678012), research project of the Ministry of Education, Science and Technological Development, Republic of Serbia (project III 46005) and Provincial Secretariat for Higher Education and Scientific Research of the Autonomous Province of Vojvodina through the project "Application of novel and conventional processes for removal of most common contaminants, mycotoxins and salmonella, in order to produce safe animal feed inthe territory of AP Vojvodina", Project No. 114-451-2505/2016-01.

Acknowledgments: We thank David Edmunds and Peter Cotty for their assistance.

Conflicts of Interest: The authors declare no conflicts of interest.

References

1. Cotty, P.J.; Bayman, P.; Egel, D.S.; Elias, K.S. Agriculture, aflatoxins and *Aspergillus*. In *The Genus Aspergillus: From Taxonomy and Genetics to Industrial Application*; Powel, K.A., Renwick, A., Peverdy, J.F., Eds.; Plenum Press: New York, NY, USA, 1994; pp. 1–27.
2. Bandyopadhyay, R.; Ortega-Beltran, A.; Akande, A.; Mutegi, C.; Atehnkeng, J.; Kaptoge, L.; Senghor, A.L.; Adhikari, B.N.; Cotty, P.J. Biological control of aflatoxins in Africa: Current status and potential challenges in the face of climate changes. *World Mycotoxin J.* **2016**, 1–20. [CrossRef]
3. Liu, Y.; Wu, F. Global burden of aflatoxin-induced hepatocellular carcinoma: A risk assessment. *Environ. Health Perspect.* **2010**, *118*, 818–824. [CrossRef] [PubMed]
4. Wu, F. Global impacts of aflatoxin in maize: Trade and human health. *World Mycotoxin J.* **2015**, *8*, 137–142. [CrossRef]
5. International Agency for Research on Cancer (IARC). Aflatoxins. In *Monograph on the Evaluation of Carcinogenic Risks to Humans. Some Traditional Herbal Medicines, some Mycotoxins, Naphthalene and Styrene*; IARC: Lyon, France, 2002; Volume 82, pp. 171–300.
6. Payne, A.G.; Yu, J. Ecology, development and gene regulation in *Aspergillus flavus*. In *Aspergillus: Molecular Biology and Genomics*; Machida, M., Gomi, K., Eds.; Caister Academic Press: Norfolk, UK, 2010.
7. Mehl, H.L.; Jaime, R.; Callicott, K.A.; Probst, C.; Garber, N.P.; Ortega-Beltran, A.; Grubisha, L.C.; Cotty, P.J. *Aspergillus flavus* diversity on crops and in the environment can be exploited to reduce aflatoxin exposure and improve health. *Ann. N. Y. Acad. Sci.* **2012**, *1273*, 7–17. [CrossRef] [PubMed]
8. Williams, W.P. Breeding for resistance to aflatoxin accumulation in maize. *Mycotoxin Res.* **2006**, *22*, 27–32. [CrossRef] [PubMed]

9. Cotty, P.J. Cottonseed losses and mycotoxins. In *Compendium of Cotton Diseases*; Kirkpatrick, T.L., Rothrock, C.S., Eds.; The American Phytopathological Society: Saint Paul, MN, USA, 2001; pp. 9–13.

10. Cotty, P.J.; Mellon, J.E. Ecology of aflatoxin-producing fungi and biocontrol of aflatoxin contamination. *Mycotoxin Res.* **2006**, *22*, 110–117. [CrossRef]

11. Battilani, P.; Toscano, P.; Van der Fels-Klerx, H.J.; Moretti, A.; Leggieri, M.C.; Brera, C.; Robinson, T. Aflatoxin B1 contamination in maize in Europe increases due to climate change. *Sci. Rep.* **2016**, *6*, 24328. [CrossRef]

12. Lević, J.; Gošić-Dondo, S.; Ivanović, D.; Stanković, S.; Krnjaja, V.; Bočarov-Stančić, A.; Stepanić, A. An outbreak of *Aspergillus* species in response to environmental conditions in Serbia. *Pestic. Phytomed.* **2013**, *28*, 167–179. [CrossRef]

13. Kos, J.; Mastilović, J.; Janić Hajnal, E.; Šarić, B. Natural occurrence of aflatoxins in maize harvested in Serbia during 2009–2012. *Food Control* **2013**, *34*, 31–34. [CrossRef]

14. Cotty, P.J.; Bayman, P. Competitive exclusion of a toxigenic strain of *Aspergillus flavus* by an atoxigenic strain. *Phytopathology* **1993**, *83*, 1283–1287. [CrossRef]

15. Cotty, P.J. Biocompetitive exclusion of toxigenic fungi. In *The Mycotoxin Factbook*; Barug, D., Bhatnagar, D., Van Egmond, H.P., Van der Kamp, J.W., Van Osenbruggen, W.A., Visconti, A., Eds.; Wageningen Academic Publishers: Wageningen, The Netherlands, 2006; pp. 179–197.

16. Atehnkeng, J.; Ojiambo, P.S.; Ikotun, T.; Sikora, R.A.; Cotty, P.J.; Bandyopadhyay, R. Evaluation of atoxigenic isolates of *Aspergillus flavus* as potential biocontrol agents for aflatoxin in maize. *Food Addit. Contam.* **2008**, *25*, 1264–1271. [CrossRef] [PubMed]

17. Dorner, J.W. Biological control of aflatoxin contamination in corn using a nontoxigenic strain of *Aspergillus flavus*. *J. Food Prot.* **2009**, *72*, 801–804. [CrossRef] [PubMed]

18. Probst, C.; Bandyopadhyay, R.; Price, L.E.; Cotty, P.J. Identification of atoxigenic *Aspergillus flavus* isolates to reduce aflatoxin contamination of maize in Kenya. *Plant Dis.* **2011**, *95*, 212–218. [CrossRef] [PubMed]

19. Callicott, K.A.; Cotty, P.J. Method for monitoring deletions in the aflatoxin biosynthesis gene cluster of *Aspergillus flavus* with multiplex PCR. *Lett. Appl. Microbiol.* **2014**, *60*, 60–65. [CrossRef]

20. Chang, P.; Horn, B.W.; Dorner, J.W. Sequence breakpoints in the aflatoxin biosynthesis gene cluster and flanking regions in nonaflatoxigenic *Aspergillus flavus* isolates. *Fungal Genet. Biol.* **2005**, *42*, 914–923. [CrossRef]

21. Yin, Y.; Lou, T.; Yan, L.; Michailides, T.; Ma, Z. Molecular characterization of toxigenic and atoxigenic *Aspergillus flavus* isolates, collected from peanut fields in China. *J. Appl. Microbiol.* **2009**, *107*, 1857–1865. [CrossRef]

22. Donner, M.; Atehnkeng, J.; Sikora, R.A.; Bandyopadhyay, R.; Cotty, P.J. Molecular characterization of atoxigenic strains for biological control of aflatoxins in Nigeria. *Food Addit. Contam.* **2010**, *27*, 576–590. [CrossRef]

23. Adhikari, B.N.; Bandyopadhyay, R.; Cotty, P.J. Degeneration of aflatoxin gene clusters in *Aspergillus flavus* from Africa and North America. *AMB Express* **2016**, *6*, 62. [CrossRef]

24. Mukanga, M.; Derera, J.; Tongoona, P.; Laing, M.D. A survey of pre-harvest ear rot diseases of maize and associated mycotoxins in south and central Zambia. *Int. J. Food Microbiol.* **2010**, *141*, 213–221. [CrossRef]

25. Williams, W.P.; Ozkan, S.; Ankala, A.; Windham, G.L. Ear rot, aflatoxin accumulation, and fungal biomass in maize after inoculation with *Aspergillus flavus*. *Field Crops Res.* **2011**, *120*, 196–200. [CrossRef]

26. Dorner, J.W.; Cole, R.J.; Wicklow, D.T. Aflatoxin reduction in corn through field application of competitive fungi. *J. Food Prot.* **1999**, *62*, 650–656. [CrossRef] [PubMed]

27. Atehnkeng, J.; Oijambo, P.S.; Cotty, P.J.; Bandyopadhyay, R. Field efficacy of a mixture of atoxigenic *Aspergillus flavus* Link: Fr vegetative compatibility groups in preventing aflatoxin contamination in maize (*Zea mays* L.). *Biol. Control* **2014**, *72*, 62–70. [CrossRef]

28. Abbas, H.K.; Zablotowicz, R.M.; Bruns, H.A.; Abel, C.A. Biocontrol of aflatoxin in corn by inoculation with non-aflatoxigenic *Aspergillus flavus* isolates. *Biocontrol Sci. Technol.* **2006**, *16*, 437–449. [CrossRef]

29. Payne, G.A.; Cassel, D.K.; Adkins, C.R. Reduction of aflatoxin contamination in corn by irrigation and tillage. *Phytopathology* **1986**, *76*, 679–684. [CrossRef]

30. Bruns, H.A. Controlling aflatoxin and fumonisin in maize by crop management. *J. Toxicol.* **2003**, *22*, 153–173. [CrossRef]

31. Guo, B.; Chen, Z.; Lee, R.D.; Scully, B.T. Drought stress and preharvest aflatoxin contamination in agricultural commodity: Genetics, genomics and proteomics. *J. Integr. Plant Biol.* **2008**, *50*, 1281–1291. [CrossRef]

32. Chauhan, Y.S.; Wright, G.C.; Rachaputi, N.C. Modelling climatic risks of aflatoxin contamination in maize. *Aust. J. Exp. Agric.* **2008**, *48*, 358–366. [CrossRef]

33. Cotty, P.J. Virulence and cultural characteristics of two *Aspergillus flavus* strains pathogenic on cotton. *Phytopathology* **1989**, *79*, 808–814. [CrossRef]

34. Barošević, T.; Bagi, F.; Budakov, D.; Kocsubé, S.; Varga, J.; Grahovac, M.; Stojšin, V. Molecular and morphological identification of *Aspergillus* species on corn seeds. In Proceedings of the III International Congress Food Technology, Quality and Safety, Novi Sad, Serbia, 25–27 October 2016; University of Novi Sad, Institute of Food Technology: Novi Sad, Serbia, 2016; pp. 365–371.

35. Brody, J.R.; Kern, S.E. Sodium boric acid: A Trisfree, cooler conductive medium for DNA electrophoresis. *Biotechniques* **2004**, *36*, 214–216. [CrossRef]

36. Garber, N.P.; Ortega-Beltran, A.; Barker, G.; Probst, C.; Callicott, K.A.; Jaime, R.; Mehl, H.L.; Cotty, P.J. *Brief Protocols for Research on Management of Aflatoxin-Producing Fungi*, 1st ed.; School of Plant Sciences, The University of Arizona: Tucson, AZ, USA, 2012; pp. 95–100.

37. Cotty, P.J.; University of Arizona, Tucson, AZ, USA. *Personal communication*, 2016.

38. Reid, L.M.; Hamilton, R.E.; Mather, D.E. *Screening Maize for Resistance to Gibberella Ear Rot*; Technical Bulletin; Agriculture and Agri-Food Canada: Ottawa, ON, Canada, 1996; p. 62.

39. European Commission No 401/2006 of 23 February 2006. Laying down the methods of sampling and analysis for the official control of the levels of mycotoxins in foodstuffs. *Off. J. Eur. Union L.* **2006**, *70*, 12–34.

40. Kebede, H.; Abbas, H.K.; Fisher, D.K.; Bellaloui, N. Relationship between aflatoxin contamination and physiological responses of corn plants under drought and heat stress. *Toxins* **2012**, *4*, 1385–1403. [CrossRef] [PubMed]

41. Republic Hydrometeorological Service of Serbia. Multiannual Average of Meteorological Parameters (1981–2010). Available online: http://www.hidmet.gov.rs/ciril/meteorologija/klimatologija_srednjaci.php (accessed on 5 November 2019).

Article

Commercial Biocontrol Agents Reveal Contrasting Comportments Against Two Mycotoxigenic Fungi in Cereals: *Fusarium Graminearum* and *Fusarium Verticillioides*

Lucile Pellan [1,*], Noël Durand [1,2], Véronique Martinez [1], Angélique Fontana [1], Sabine Schorr-Galindo [1] and Caroline Strub [1]

[1] Qualisud, Univ Montpellier, CIRAD, Montpellier SupAgro, Univ d'Avignon, Univ de La Réunion, Montpellier, France; noel.durand@cirad.fr (N.D.); veronique.martinez@umontpellier.fr (V.M.); angelique.fontana@umontpellier.fr (A.F.); sabine.galindo@umontpellier.fr (S.S.-G.); caroline.strub@umontpellier.fr (C.S.)

[2] CIRAD, UMR Qualisud, F-34398 Montpellier, France

* Correspondence: lucile.pellan@umontpellier.fr

Received: 7 February 2020; Accepted: 24 February 2020; Published: 29 February 2020

Abstract: The aim of this study was to investigate the impact of commercialized biological control agents (BCAs) against two major mycotoxigenic fungi in cereals, *Fusarium graminearum* and *Fusarium verticillioides*, which are trichothecene and fumonisin producers, respectively. With these objectives in mind, three commercial BCAs were selected with contrasting uses and microorganism types (*T. asperellum*, *S. griseoviridis*, *P. oligandrum*) and a culture medium was identified to develop an optimized dual culture bioassay method. Their comportment was examined in dual culture bioassay in vitro with both fusaria to determine growth and mycotoxin production kinetics. Antagonist activity and variable levels or patterns of mycotoxinogenesis inhibition were observed depending on the microorganism type of BCA or on the culture conditions (e.g., different nutritional sources), suggesting that contrasting biocontrol mechanisms are involved. *S. griseoviridis* leads to a growth inhibition zone where the pathogen mycelium structure is altered, suggesting the diffusion of antimicrobial compounds. In contrast, *T. asperellum* and *P. oligandrum* are able to grow faster than the pathogen. *T. asperellum* showed the capacity to degrade pathogenic mycelia, involving chitinolytic activities. In dual culture bioassay with *F. graminearum*, this BCA reduced the growth and mycotoxin concentration by 48% and 72%, respectively, and by 78% and 72% in dual culture bioassay against *F. verticillioides*. *P. oligandrum* progressed over the pathogen colony, suggesting a close type of interaction such as mycoparasitism, as confirmed by microscopic observation. In dual culture bioassay with *F. graminearum*, *P. oligandrum* reduced the growth and mycotoxin concentration by 79% and 93%, respectively. In the dual culture bioassay with *F. verticillioides*, *P. oligandrum* reduced the growth and mycotoxin concentration by 49% and 56%, respectively. In vitro dual culture bioassay with different culture media as well as the nutritional phenotyping of different microorganisms made it possible to explore the path of nutritional competition in order to explain part of the observed inhibition by BCAs.

Keywords: antagonistic agents; in vitro dual culture bioassay; mycotoxins; nutritional competition

Key Contribution: Complementary and integrative approaches to characterize the impact of three distinct commercial biocontrol agents on the pathogenic capacities of two mycotoxingenic fungi, including the monitoring of growth, spore production, mycotoxin production or nutritional profile.

---

1. Introduction

Many plant pathogens endanger agriculture, particularly in cereal production. Some of these phytopathogens can produce mycotoxins and have multi-destructive effects through reducing the yield and nutritional properties of grain, and can also affect its health-related characteristics. Among mycotoxigenic fungi, the genus *Fusarium* is the most prevalent and represents a significant risk [1].

Mycotoxins are small toxic molecules resulting from the secondary metabolism of these fungi. *Fusarium graminearum* is able to produce deoxynivalenol (DON) and its derivatives 15 and 3-ADON. *Fusarium verticillioides* mainly produces fumonisins B_1 and fumonisins B_2 (FB$_1$ and FB$_2$). They are ubiquitous in agricultural production, and cereal crops are the primary source of consumer exposure to mycotoxins [2]. Mycotoxins are quite stable molecules which are very difficult to remove or degrade, and are found along the food chain while retaining their toxic properties. They can cause acute and chronic intoxication in both humans and animals depending on different factors (intake levels, duration of exposure, toxin type, mechanisms of action, metabolism, and defense mechanisms) [3]. The symptoms can be very variable depending on organisms considered, but can range from food poisoning or gastric-intestinal problems to suppression of the immune system, the induction of cancer, or death [4,5]. Sometimes the combination within trichothecenes/fumonisins or with other mycotoxins such as ochratoxins or aflatoxins can produce a synergistic toxic effect [6,7]. Given their high toxicity, DON and fumonisins have been regulated by the European Union and the maximum recommended levels of mycotoxins are 750 µg kg^{-1} and 1000 µg kg^{-1}, respectively, for processed cereal intended for human consumption (Commission Regulation (EC) No 401/2006). Economically, these contaminants hamper international trade and significantly affect the world economy. Because their appearance can occur before or after harvesting, one of best ways to reduce the presence of mycotoxins in food and feed is to prevent their formation in the crop [8].

Agricultural production is subject to a scissor effect with, on one hand, very evolutive climatic conditions that promote microbial development and, on the other hand, the political determination to reduce the use of phyto-chemical products. The adverse effects of plant protection products on environment and on human and animal health have encouraged the European Union to promote the search for alternative and environmentally friendly solutions such as integrated pest protection and the use of biological control agents (BCAs) [9]. Non-pathogenic, they can prevent or reduce the progression of these fungi through different mechanisms. These mechanisms are very diverse, direct or indirect, and can be used together or alone depending on the biocontrol agent. The main direct mechanisms may be related to antibiosis, competition, parasitism, or toxin biotransformation [10]. Antibiosis includes the production of secondary metabolites like antibiotics, cell wall-degrading enzymes, and molecules that can suppress growth or kill pathogens [11–13]. Competition can occur when two or more fungi are present simultaneously. Microorganisms can then enter into nutritional competition when they need the same essential nutrient, or into spatial competition when they occupy the same limited space [14,15]. Parasitism is a direct attack of pathogen by the BCA that will degrade or consume the pathogen host until it dies [16]. Some mechanisms can be indirect. Colonization of the plant by beneficial microorganisms can trigger local or systemic defensive responses, enhancing resistance against phytopathogens [17,18].

Although the production of mycotoxins by toxigenic pathogens is of economic importance, many studies do not take this into account when studying biological control strategies. These studies are then limited to the fungicidal or fungistatic effects of BCAs, while the effect of BCAs on mycotoxin production is often neglected [19]. Moreover, most studies on the biocontrol of other plant pathogens do not consider the potential presence of several mycotoxins simultaneously in the crop, nor the side effects of BCAs on the mycotoxigenic agents present. Despite more than 70 years of intensive research and some promising results, only a few BCAs are currently available on the market, and for *Fusarium spp.* for instance in Europe, only *Pseudomonas chlororaphis* and *Pythium oligandrum* are available, as the commercial products Cerall® (Belchim Crop Protection) and Polyversum® (Biopreparáty/De Sangosse), respectively.

The lack of commercially available BCAs for mycotoxigenic fusaria may be due to numerous factors (unstable biocontrol efficacy under field conditions, strict storage and transport conditions for BCAs, complexity of registration) and more importantly a lack of knowledge about biocontrol mechanisms [20], especially concerning mycotoxins.

Indeed, a better understanding of the mechanisms that allow the biocontrol of pathogens, especially on marketed agents that have passed all the stages of formulation and registration, will make it possible to identify particularly effective mechanisms or biomarkers of mechanisms. This information could be used to optimize the use of existing BCAs but also to refine the selection of new BCAs.

The ultimate goal of the present study was to characterize the impact of commercial biocontrol agent on life traits of mycotoxigenic fungi (growth and mycotoxinogenesis). Therefore, the objectives were (1) to develop optimal methods of dual culture bioassay, allowing the analysis of the interaction between pathogens and BCAs; (2) to identify effect of BCAs on growth and mycotoxins of *Fusarium graminearum* and *Fusarium verticillioides*; and (3) to propose a hypothesis on the modes of action of BCAs against mycotoxigenic pathogens. With these objectives in mind, three commercial BCAs were selected with contrasting uses and microorganism types (*T. asperellum*, *S. griseoviridis*, *P. oligandrum*) and studied with in vitro dual culture bioassay against *F. graminearum* and *F. verticillioides* with an optimized method promoting mycotoxin production. Variable levels of mycotoxinogenesis and growth reduction were observed depending on the microorganism type of BCAs or on the culture conditions (e.g., different nutritional sources), suggesting contrasting biocontrol mechanisms. Macroscopic and microscopic observation of interactions support this hypothesis, finding various structural formations of BCA. Nutritional phenotyping was used to detect the possibility of nutritional competition between microorganisms, and to identify the trophic requirements of BCAs.

2. Results

To permit the interaction between BCAs and pathogens during dual culture bioassay, the first step was to select appropriate media to allow the development of each microorganism's pairing (BCA/pathogen). BCA/pathogen selection criteria and descriptions are presented in Section 5, Material and Methods. Then, to characterize this interaction on selected media, integrative approaches were used. Microscopic and macroscopic observations were undertaken to identify specific comportment of BCAs and their effects on *Fusarium* physiology. The BCA treatment's impact on growth, global mycotoxins, and specific mycotoxin production kinetics of pathogens was analysed. Finally, the impact of differential nutritional resources during dual culture bioassay was analysed and linked with the nutrient profiling of all microorganisms.

2.1. Comportments of Microorganisms and Culture Media Selection

After 8 days of incubation, all microorganisms were able to grow on all different media, with the exception of Polyversum® (Poly), which was unable to grow on Czapeck agar (CZA) synthetic medium (Figure 1). Mycostop® (Myco), which has the most limited growth (area under the growth curve, AUGC < 1.85×10^2), was an excellent spore producer (spores > 3.85×10^3), especially on International *Streptomyces* Project Medium 2 (ISP2). Xedavir® (Xeda) was minimally affected by the culture medium variation. Polyversum® (Poly), the oomycete BCA, was characterized solely on growth capacity because oospore obtention was a long and complex process which is difficult to achieve, and to obtain oospores under these specific in vitro conditions. *F. graminearum* produced a low quantity of spores affected by medium variation. For both pathogens, global mycotoxin production was a very discriminant variable for medium selection. The culture media on which pathogens *F. graminearum* and *F. verticillioides* produced the most mycotoxins were respectively CZA and PDA.

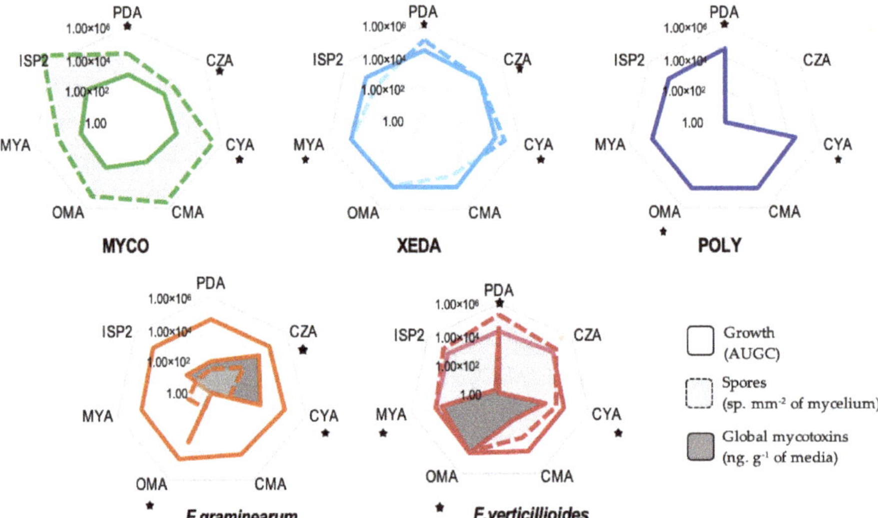

Figure 1. Comportment of all micro-organisms (commercialized biological control agents (BCAs) and pathogens) on different media. Stars indicate the final selected medium for each microorganism. Myco: Mycostop®, Xeda: Xedavir®, Poly: Polyversum®. AUGC: area under the growth curve, PDA: potato dextrose agar, CZA: Czapeck agar, CYA: Czapeck yeast agar, CMA: corn meal agar, OMA: oat meal agar, MYA: malt yeast agar, ISP2: International *Streptomyces* Project Medium 2 (8 days at 25 °C). Levels of global mycotoxins are given as a sum of the mycotoxins from a pathogen chemotype (*Fusarium graminearum*: deoxynivalenol (DON) + 15ADON and *Fusarium verticillioides*: fumonisins B_1 (FB_1) + fumonisins B_2 (FB_2)).

In order to promote pathogen development and mycotoxin production during interaction, the selection of dual culture bioassay media was firstly based on the mycotoxin production capacity of pathogen (Table 1). Then, the ability of BCAs for growth on this medium was checked. To observe the differential capacity of BCAs to impact mycotoxigenic fungi, one common medium (CM) for each pathogen was selected: for *F. graminearum*–BCA dual culture bioassay CYA was chosen, and for *F. verticillioides*–BCA dual culture bioassay PDA was selected. Other interesting media which did not allow all BCAs to grow were assigned to a specific dual culture bioassay duo (SM), and were selected for their ability to promote mycotoxinogenesis. These selected culture media were used to test antagonist activity and allowed identification of the impact of BCAs on pathogenic growth and mycotoxin production. The use of different culture media enabled the characterization of the impact of nutrient resources on the dual culture bioassay of microorganisms.

Table 1. Selected media for dual culture bioassay. CM: common medium for all BCAs confrontation of concerned pathogen; SM: Specific media for each BCA-pathogen confrontation of concerned pathogen. Myco: Mycostop®, Xeda: Xedavir®, Poly: Polyversum®.

Selected Media for Dual Culture Bioassay (Pathogen–BCA)				
Pathogens	***F. Graminearum***		***F. Verticillioides***	
BCA	**CM**	**SM**	**CM**	**SM**
Myco	CYA	CZA	PDA	CYA
Xeda	CYA	CZA	PDA	MYA
Poly	CYA	OMA	PDA	OMA

2.2. Macroscopic and Microscopic Interaction

During the dual culture bioassay test, the macroscopic observation of plates (7 days after inoculation) revealed differential comportments between microorganisms (Figure 2). For both pathogens, mycelial development was limited by the presence of all BCAs compared to the control. Myco formed a very small colony related to pathogens, but an inhibition zone of pathogen growth was observed (especially in dual culture bioassay with *F. verticillioides*). With respect to pathogens, accumulation of pigment was observed in the dual culture bioassay zone (on the reverse side of Petri dish, not shown). The Xeda colony grew as fast as the pathogen and filled the free space in the Petri dish. Once again, accumulation of pathogenic pigment in the confrontation zone could be observed on the reverse side of Petri dishes. The Xeda colony could exude fine black droplets on its surface solely during confrontation. Poly was able to grow faster than the pathogen (especially faster than *F. graminearum*) and fill the free space in Petri dish. During co-culture, it was able to progress on both the pathogenic colony and inhibit the homogenous and natural pink pigmentation of *F. graminearum* completely.

At the microscopic scale, *F. graminearum* alone produced rapidly long, dense, and straight hyphae, while *F. verticillioides* alone produced fine and extra fine lightly branched hyphae (Figure 2). With microscopic interactions between BCAs and pathogens photographed over time, many characteristic structures could be observed. Myco was able to cause anarchic ramifications of *F. graminearum* mycelia before contact (+72 h) and vesicle formation at the tip of pathogen hyphae when they were close (+96 h). Growth inhibition by Myco was clearly identified at the microscopic level in co-culture with *F. verticillioides*, with an early antigerminative action (+72 h) and a linear/regular growth inhibition after 96 h. Xeda induced deformation of *F. graminearum* mycelia, which could extend to the formation of loops after 96 h. The progression of this BCA face-to-face with *F. verticillioides* created a mechanical barrier that slowed the progression of the mycotoxigenic fungus. In the dual culture bioassays of both pathogens, Xeda caused a degradation of pathogenic mycelia in contact zones (+96 h). The co-culture *F. graminearum*–Polyversum showed dispersed contact structures. The BCA was able to pass through the mycelium of the pathogen, and induced formation of lysed wall hyphae. It adopted a completely different comportment in the dual culture bioassay against *F. verticillioides* and induced the transient formation of micro-vesicles at the extremity of pathogenic hyphae (between 24 and 72 h of dual culture bioassay). When it was close to the pathogen (+96 h), it deployed long mycelial filaments to colonize the environment.

2.3. BCA Impact on Growth and Mycotoxin Production

During the antagonistic in vitro test, the growth and the mycotoxin production of both pathogens in confrontation against the three BCAs were monitored over 12 days. Classic batch curves were obtained for pathogen growth and major mycotoxin production (latent phase, exponential phase, stationary phase). The area under growth/global mycotoxins curve (respectively AUGC and AURMC) was calculated (Figure 3). All BCA treatments impacted growth of *F. graminearum*, especially Xeda and Poly, causing growth reductions of respectively 48% and 79%.

In addition to reducing the pathogens mycelial development, Xeda and Poly had a direct action on global DON production, with reductions in mycotoxin concentrations of 73% and 93% respectively. *F. graminearum* 15-ADON production was affected and complete inhibition of 15-ADON peak production (8 days after inoculation) was observed with all BCA treatments (see Figure A2). Dual culture bioassays with *F. verticillioides* showed a completely different response profile to BCAs. Even if the growth reduction of pathogen was significant face-to-face with Myco (24%), the specific production was lightly stimulated. It was mainly with the reduction of pathogen growth that Xeda restricted fumonisin production, with the most important inhibition levels (78%) among all BCA tested.

In summary, Poly was more efficient in the *F. graminearum* dual culture bioassay and Xeda was more efficient in the *F. verticillioides* dual culture bioassay using common media (CM).

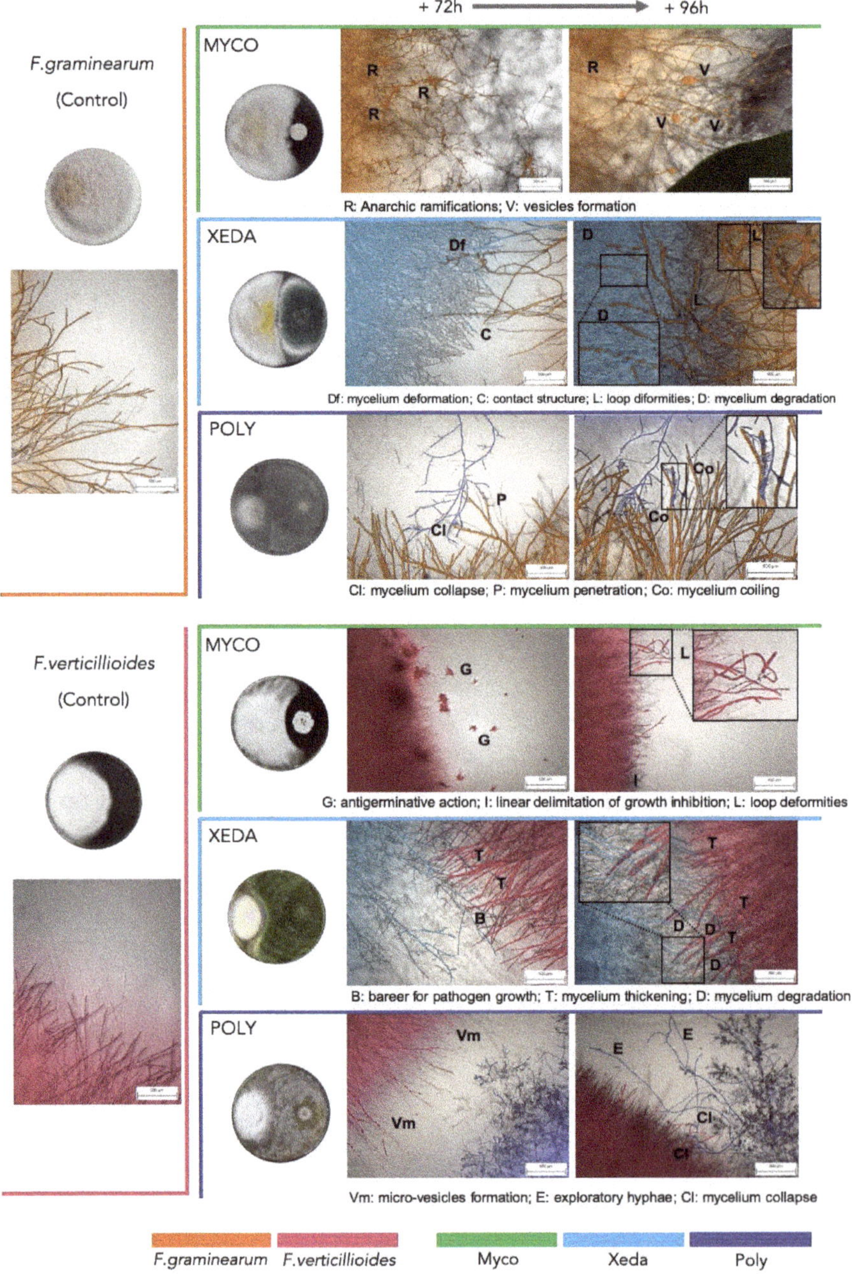

Figure 2. Macroscopic and microscopic interactions during pathogen–BCA dual culture bioassay. The *F. graminearum* and *F. verticillioides* control and confrontation plates respectively had CYA and PDA media (CM, 7 days after inoculation). The same media were used for the slide confrontation (+72/96 h after inoculation) indicated at the top of the figure. Colors indicate the type of microorganisms, and letters indicate specific structures. Myco: Mycostop®, Xeda: Xedavir®, Poly: Polyversum®.

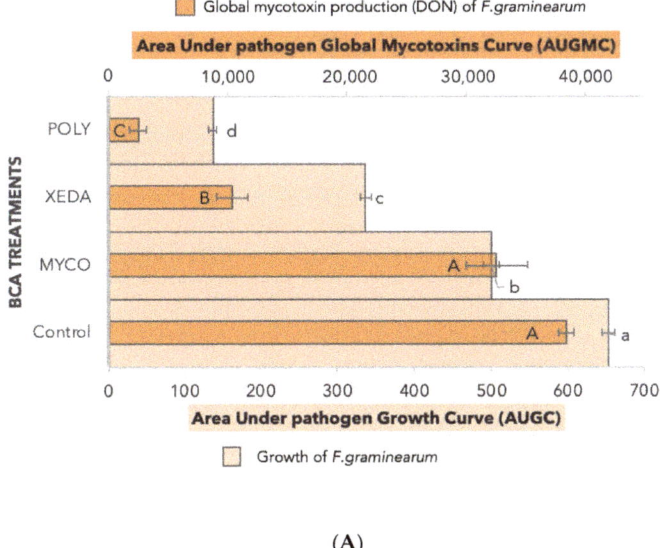

(A)

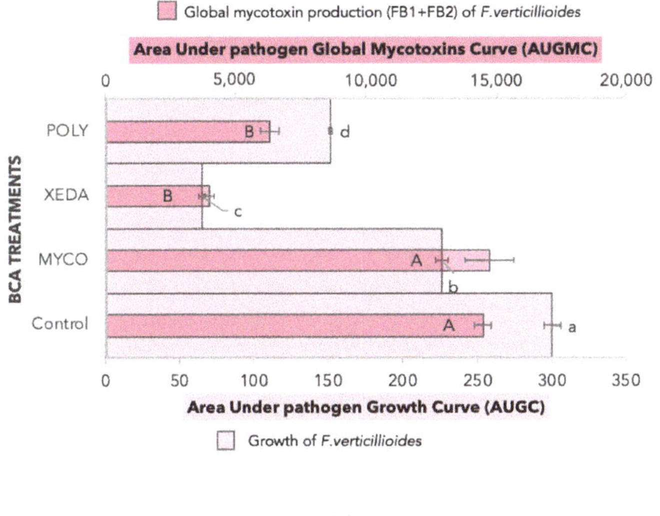

(B)

Figure 3. Comparative effects of BCA treatments on the evolution of pathogens. Dual culture bioassays for all BCAs against one pathogen were performed on common medium (CM). (**A**) *F. graminearum*, on CYA, in orange; (**B**) *F. verticillioides*, on PDA, in pink, for 12 days. Myco: Mycostop®, Xeda: Xedavir®, Poly: Polyversum®. Growth (pastel colors, bottom axis) and mycotoxin production levels (pastel colors, top axis) are expressed in the area under growth/global mycotoxin curves (respectively AUGC and AUGMC). Original kinetic curves are available in Figure A2. ANOVA test, Growth comparisons (a, b, c, d) and Mycotoxin comparisons (A, B, C). p-value < 0.05. $FB_1 + FB_2$ are represented together because the quantity of FB2 was not significant.

2.4. Impact of Media Variation on Pathogenic Factors during Dual Culture Bioassay (Pathogens–BCAs)

Principal component analysis (PCA) analysis was performed in order to visualize global inhibition profiles of pathogens depending on BCA treatment, but also to assess the impact of culture medium on those inhibition profiles (Figure 4). For both biplots, the variables are well-projected and equivalently projected, and the two first components explained 90.7% and 94.8% of data distribution for *F. graminearum* (Figure 4A) and *F. verticillioides* (Figure 4B), respectively. The growth factor mainly contributed to Dim2, mycotoxin factors mainly contributed to Dim1, and all variables were in correlation. With all culture media combined, Myco and Control samples were regrouped in both graphs in the positive part of biplot. However, culture media influenced the separation of samples on CZA (*F. graminearum* control or confronted with Myco). Control samples were correlated with DON and 15-ADON factors, in contrast to Myco treatment samples, on the other side of Dim2. In the *F. graminearum* bi-plot (Figure 4A), Xeda and Poly samples are regrouped in the bottom-left quarter of graph, suggesting strong opposition to the studied pathogen factor. For the Poly-confronted samples, an increase of mycotoxin inhibition was observed in OMA, highlighting the important effect of culture media on dual culture bioassay, illustrated here for *F. graminearum*. Global antagonist activity of Xeda against *F. graminearum* in the CYA medium was higher than on CZA. On the *F. verticillioides* bi-plot (Figure 4B), these two BCAs (Xeda and Poly) caused the most antagonistic activity. However, against this pathogen, Xeda showed the strongest growth and mycotoxin inhibition. MYA-confronted samples were further from the specific mycotoxin production factor than PDA-confronted samples, indicating a higher specific mycotoxin reduction in this culture medium. Poly was able to inhibit pathogenic factors in both culture media, but with OMA the separation between control and confronted samples was greater, suggesting a stimulation of antagonistic capacities on this culture medium.

2.5. Nutrient Profiling and Nutritional Competition

Given the observed medium impact, nutrient profiling of all microorganisms was characterized using phenotype microarray, and a comparison between microorganisms was performed to identify potential nutritional competition. Along 138 h of incubation, a contrasting capacity to consume the different nutritional resources appeared. As shown in Figure 5, globally, two clusters of microorganisms can be distinguished based on the capacity to metabolize C sources (Figure 5A), with separate microorganisms in two clusters: one with Poly and *F. graminearum*, which were able to consume less C sources than others (high trophic requirements), and one including *F. verticillioides*, Xeda, and Myco, which could consume a broad range of C sources (low trophic requirements). Some C compounds strongly stimulated the growth of microorganisms, such as tyramine/glucose, N-acetylglucosamine/sucrose, threonine/tartaric acid, glycerol/uridine, and proline/sucrose, respectively, for Myco, Xeda, Poly, *F. graminearum*, and *F. verticillioides*. Sucrose was therefore the carbohydrate allowing the strongest growth of two antagonistic microorganisms (Xeda and *F. verticillioides*). On other hand, threonine is a compound that is consumed only by the Poly BCA and inhibits the growth of other microorganisms, including the two mycotoxigenic pathogens. Concerning the aptitude of microorganisms to degrade nitrogen sources (Figure 5B), two different clusters were formed: one with of Poly, Xeda, and Myco, which were able to consume less N sources than others (specialized); and one including *F. graminearum* and *F. verticillioides*, which could consume a broad range of N sources (unspecialized). This clustering was in accordance with the type of microorganism (BCAs or pathogens), and Poly has a nitrogen consumption profile which is in complete contrast to the *F. verticillioides* profile. Some N nutrients increase the growth of particular microorganisms: histidine/lysine, uric acid/serine, xanthine/L-cysteine, uric acid/allantoin, and serine /L-pyroglutamic acid, respectively, for Myco, Xeda, Poly, *F. graminearum*, and *F. verticillioides*. Preferential nitrogen sources of Xeda were also the preferential nitrogen sources of the two pathogens.

Nutrient consumption and the resulting growth capacity were difficult to relate to the synergistic effects that could be observed in the dual culture bioassay on complex synthetic media. Nevertheless, results show that Poly, because of its high trophic requirement, has more difficulty consuming nutrients than pathogens. This observation is confirmed with regard to the major elements that comprised the culture media of the dual culture bioassay (the two CM pathogens PDA and CYA, and the SM BCAs OMA), even on OMA, where it showed very strong antagonistic capacities during dual culture bioassay. Xeda had better biocontrol activity against *F. verticillioides* on MYA (global mycotoxin inhibition), and against *F. graminearum* on CYA (growth and global mycotoxin inhibition). The MYA medium was composed of fructose and mannitol, which the pathogen/BCA duo can degrade in equivalent proportions. CYA medium contains sucrose and nitrate, two compounds for which Xeda has a stronger affinity than *F. graminearum*.

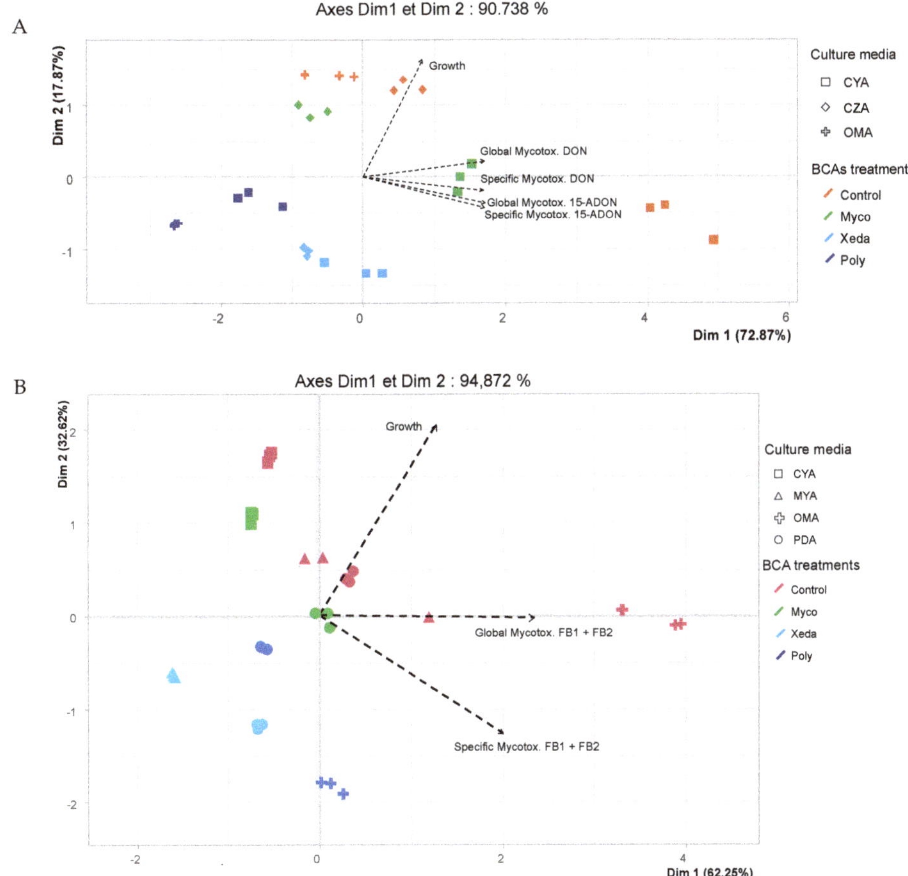

Figure 4. Principal component analysis (PCA) bi-plot with respect to the growth and mycotoxin production data set (—) for the two considered pathogens, (**A**) *F. graminearum* and (**B**) *F. verticillioides*, obtained after BCA confrontation (indicated by colors) in different selected culture media (indicated by forms). Myco: Mycostop®, Xeda: Xedavir®, Poly: Polyversum®.

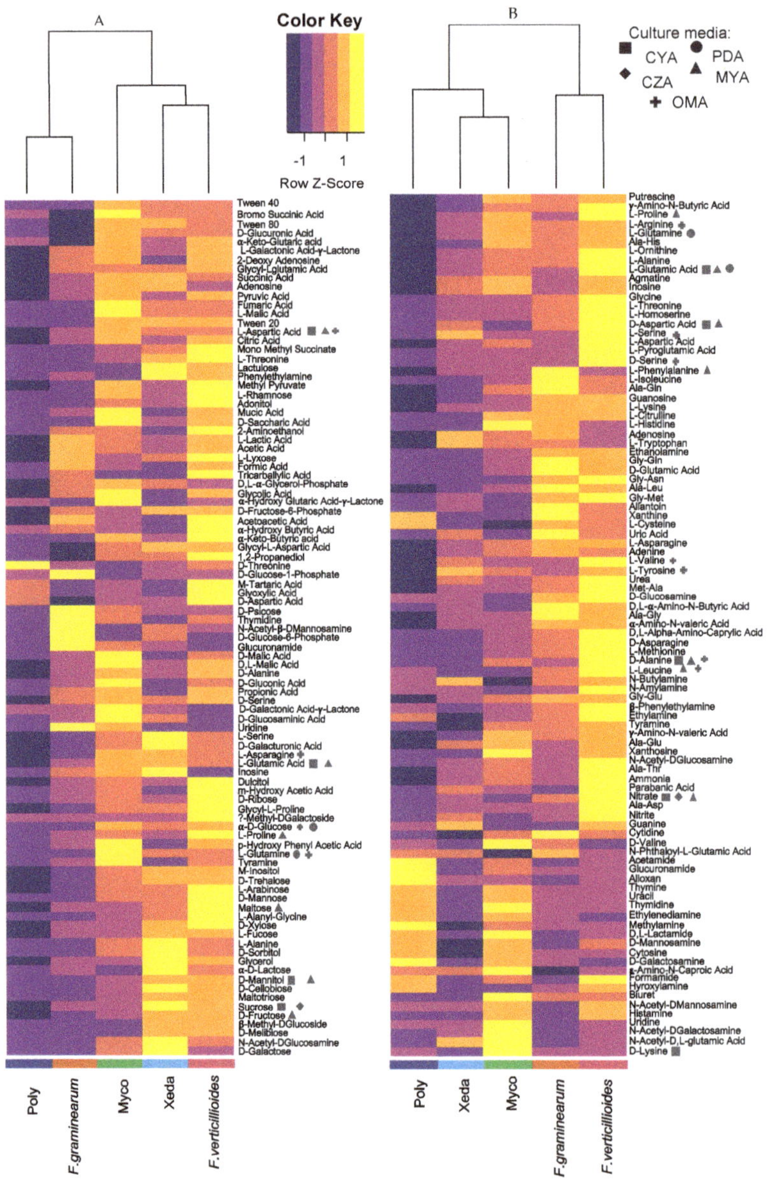

Figure 5. Growth induction of microorganisms during growth on 95 carbon sources (PM1) (**A**) or 95 nitrogen sources (PM3) (**B**). Both panels show optical density measurements over a 138-h time course (i.e., area under the growth curves, AUGCs). Color keys of the heat map (from purple to yellow) are expressed for one nutrient in comparison with other microorganisms (row comparison). Micro-organisms are indicated by colors at the end of each column. Myco: Mycostop®, Xeda: Xedavir®, Poly: Polyversum®. Major elements included in the composition of selected media are indicated by forms next to the concerned nutrient. Medium compositions are available in Figure A1.

3. Discussion

Based on the presented results, the development of dual culture bioassay methods and in particular the choice of culture media should be considered during in vitro studies between mycotoxigenic pathogens and BCAs. Globally, complex culture media selected for the affinity that the microorganisms have for them allow good growth of all microorganisms, with the exception of Poly which does not grow on CZA, a minimum medium. However, the mycotoxinogenesis of pathogens is strongly influenced by the type of media, as shown by our results and as widely reported in the literature [21,22]. A study conducted by Gardiner et al. [22] on the nutritional profile of *F. graminearum* identified the same type of profile, with preferential consumption of the same nutrient as in our results. It is therefore necessary to select culture media that will promote production of this mycotoxin in order to identify BCAs that will have a considerable impact even during high production. Most of the in vitro dual culture bioassays between pathogenic fungi and BCAs are performed on PDA [23,24], which does not stimulate the production of tricothecenes by *F. graminearum*. The use of two media per BCA/pathogen coupling allows the observation of whether nutritional conditions impact the inhibition caused by BCAs. The high mycotoxigenic capacity of the strains was verified. Many variations can be observed between the two pathogens. In natural conditions, they originate from different environments, as *F. graminearum* is isolated from wheat spikes while *F. verticillioides* is isolated from maize kernels. In addition, the pathogens produce distinct types of mycotoxins (trichothecene for *F. graminearum* and fumonisins for *F. verticillioides*) and their production levels are also different. According to the analyses that we performed on both pathogenic strains, the maximum production of fumonisin and trichothecene was about 5000 and 15,000 ng g^{-1} of media, respectively. Moreover, their colonization strategies are in contrast: *F. graminearum* grew rapidly while producing fewer spores, unlike *F. verticillioides* for which expansion was slower, but it compensated with a substantial production of spores, allowing it to colonize the environment in a comparable way. For these reasons, most of the analyses are carried out independently.

In dual culture bioassay with pathogens, Myco did not have time to develop, which did not allow it to compete spatially with pathogens. The halos of inhibition and the absence of contact during the observation period as well as the deformation of the pathogenic hyphae of *F. graminearum* suggest the synthesis of diffusible antimicrobial compounds by the BCA. Actinomycetes, and particularly *Streptomyces* species, are well known for their production of a wide spectrum of antibiotics [25,26]. In dual culture bioassay with *F. verticillioides*, they were even able to remotely prevent germination of pathogenic spores and completely inhibit pathogen progression. Several compounds were emitted by *Streptomyces spp.* and identified (e.g., as methyl vinyl ketone) for their ability to inhibit the spore germination of phytopathogens [27,28]. The appearance of vesicles at the tips of the hyphae of *F. graminearum* may be related to the synthesis of mycotoxins. Some studies have found that mycotoxins, notably in *F. graminearum*, were synthesized in toxisomes, a kind of vesicle excreted out of the cells and at the tip of hyphae [29,30]. The pathogen's perception of BCA may stimulate this production. Based on the results of the antagonistic tests, Myco tends to stimulate specific mycotoxin production despite significant inhibition of the growth of both pathogens. Therefore, it is essential to test the ability of BCAs to impact mycotoxin production and not only growth. Its ability to degrade many sources of C allows it to grow under many conditions, which could make it a preventive treatment. However, to effectively control pathogens, the ratio of inoculum from Myco must be superior than the quantity of pathogens.

Xedavir adopts a completely different strategy. With good colonization capacities, can compete with the development of pathogens. It can form a mechanical barrier that physically blocks the development of pathogenic fungi and is also be able to establish connections between its hyphae and those of pathogens through the formation of haustaurium [31]. When in contact with pathogens, it can also synthesize compounds that degrade the fungal walls, probably chitinases, preventing the progression of pathogens. Numerous studies have shown this ability in different species of *Trichoderma* [32,33]. Not surprisingly, the genomes of the mycoparasitic *Trichoderma spp.* are rich in gene-encoding enzymes like chitinases and glucanases [34,35]. This wall degradation will provide a preferential substrate for further development. It will rather have a hyperparasitic behavior [36]. Its goal is to kill pathogens. Faced with *F. graminearum*, it will act directly on the specific mycotoxinogenesis independently of growth reduction. It will therefore be able to act directly on the mycotoxin biosynthesis pathways, or by biotransformation of the mycotoxins produced (transformation, degradation, binding). It would be interesting to explore this potential because recent studies have identified mycotoxin degradation capacities in other BCAs of the *Trichoderma* genus [37]. During in vitro dual culture bioassays on selected media, culture media seem to influence the impact of Xeda on pathogens. The nutritional resources of microorganisms will therefore impact the inhibition of pathogens by BCAs. One culture medium could even allow Xeda to act more drastically on specific mycotoxinogenesis (MYA–*F. verticillioides*), and thus reduced the mycotoxin concentrations of samples. These results suggest that nutritional competition phenomena could be one of the weapons used by Xeda to fight pathogens. This path is supported by the comparison of nutritional profiles of Xeda and pathogen nutritional profiles. It appears that for certain compounds, the capacities of assimilation will be equivalent or to the advantage of BCAs, notably certain culture media compounds used during the dual culture bioassays (like sucrose). Indeed, culture media that allow a better inhibition of pathogens (MYA and CYA, respectively for *F. verticillioides* and *F. graminearum*) will be considered as resources for which microorganisms are in competition. Recently, Wei and collaborators showed that root-associated bacterial communities with a clear niche overlap with the pathogen reduce pathogen invasion success, constrain pathogen growth, and have lower levels of diseased plants in greenhouse experiments [38].

Concerning interactions between Poly and fusaria pathogens, Poly is one of the most competitive BCAs despite its high nutritional requirements. It is able to proliferate very rapidly under favorable conditions. Against the two pathogens, it will be able to invade the space left free more quickly, until it grows on the colonies of the pathogen. This behavior suggests an interaction requiring close contact, like parasitism. Poly is known for its capacities to colonize fungi and oomycetes [39]. The perception of the pathogen could be done via the perception of ergosterol from pathogens. In fact, sterols have been identified first as stimulating the sexual reproduction of *Pythium* species (like Poly) [40–42]. The same stimulation effects have been observed during dual culture bioassays with fungal pathogens [43]. This ability is a considerable advantage for the invasion of Poly and its mechanisms of action against pathogens. In addition to limiting the spatial spread of the pathogen, the absence of pathogenic pigment suggests a direct interaction with the metabolism of mycotoxigenic pathogens. At the same time after inoculation, compared to Xedavir, which has a slower progression, the contact took place in a very subtle manner. It seems that the BCA did not want to be perceived by the pathogens and only few wall degradation enzymes were synthesized. In dual culture bioassay against *F. graminarum*, Poly was able to pass through the mycelium of the pathogen and induce formation of lysed wall hyphae. They could be considered as privileged spaces for nutrient sampling by the BCA, suggesting again a close-contact interaction, such as mycoparasitism. These observations are entirely consistent with another study by Charlène Faure et al., who had observed the same type of behavior by Poly, capable of crossing the mycelium of *F. graminearum* without forming any particular structure [43]. During the antagonist activity test, Poly showed very good antagonistic capacities, allowing a reduction of the growth of both pathogens. However, the impact on mycotoxin production was much stronger on *F. graminearum*, against which it was particularly effective. By integrating the impact of the growing medium, Poly was the BCA that was most sensitive to change in nutritional resources. From the beginning of the study, it was the microorganism that was the most sensitive to culture media, with an inability to produce oospores in vitro in synthetic culture media that are not enriched in sterols. In contrast, it was able to deploy its mechanisms of action during dual culture bioassays because it was stimulated by the presence of the pathogen and by the complex nutrients in the culture media. In fact, it was most effective on a culture medium composed mainly of cereals (OMA). The results showed that Poly was less adapted in terms of nutrient consumption than the pathogens on all the culture media of the dual culture bioassays, even on OMA, where it showed very strong antagonistic capacities during dual culture bioassay test. It would therefore appear that unlike Xeda, nutrient competition was not a preferential mechanism for Poly to deal with mycotoxigenic fungal pathogens. Nevertheless, it is interesting to note that threonine was a compound that was consumed only by Poly BCA and did not promote growth of other microorganisms, including the two mycotoxigenic pathogens. It was interesting to test the enrichment of commercial formulations with compounds of this type in order to boost the growth of Poly during treatment in real conditions.

The three BCAs therefore have contrasting and complementary strategies, with different capacities to inhibit the propagation of pathogens. Their behavior was identical in one aspect: all inhibited the peak production of 15-ADON produced by *F. graminearum* (see Figure A2A). 15-ADON is a precursor to the formation of DON [44]. This inhibition could therefore be a sign of an action on the biosynthesis pathway, in particular the *Tri8* gene involved in the transformation of 15-ADON into DON.

4. Conclusions

In the present study, three commercial biological control agents (BCAs) were selected and tested to evaluate their antagonistic activities against *F. graminearum* and *F. verticillioides*, two mycotoxigenic fungi. Results suggest that this three BCAs revealed contrasting effects on growth and mycotoxin production of pathogens, particularly Xedavir® against *F. verticillioides* and Polyversum® against *F. graminearum*, the two most effective combinations. In addition, several paths could be explored concerning the various observed modes of action of these BCAs, like antibiosis, parasitism, or mycotoxin degradation. It would appear that the inhibition effect of BCAs could be dependent on the nutritional condition of interaction, providing new insight into the nutritional phenotype and potential competition between micro-organisms. Given the high levels of inhibition observed, further studies on the modes of action of these BCAs could be conducted, for example on synthesis of anti-germinative compounds by Mycostop®, chitinolytic activities or by-products of microbial detoxification of mycotoxins by Xedavir®, or mycophageous action by Polyversum®.

5. Material and Methods

5.1. Micro-Organisms

5.1.1. *Fusarium* Species

F. graminearum isolate BRFM 1967 and *F. verticillioides* isolate BRFM 2251 were used (CIRM, University of Aix-Marseille, Marseille, France), and chosen for their strong ability to produce their respective mycotoxins. *F. graminearum* strain BRFM 1967, isolated from wheat plant, has a deoxynivalenol (DON/15-ADON) chemotype profile, while *F. verticillioides* BRFM 2251, isolated from maize kernels, has a fumonisins (FB_1/FB_2/FB_3) chemotype profile. All fungi were maintained on potato dextrose agar (PDA; BD Difco, Sparks, MA, USA) under paraffin oil at 4 °C and actively grown on PDA at 25 °C for 7 days for spore production.

5.1.2. Commercial Biological Control Agents (BCAs)

Three commercial biological control agents were selected for their contrasting characteristics. To date, there is little information on their action and biocontrol mechanisms on mycotoxigenic fungi (Table 2).

All strains were isolated from their commercial product with a classical microbial insulation protocol (serial dilution and multiple striation inoculation), on appropriate medium (supplemented with 0.01% of tetracycline for Polyversum only). BCAs were conserved under spore forms in glycerol solution (15%/−80 °C) and in commercial product aliquots (4 °C). The strains were actively grown on ISP4, PDA, and V8, respectively, for Mycostop, Xedavir, and Polyversum at 25 °C for 7 days for spore production. For the rest of the study, the strains isolated from commercial products were referred to using the following abbreviations: Myco for Mycostop/*S. griseoviridis*, Xeda for Xedavir/*T. asperellum*, and Poly for Polyversum/*P. oligandrum*.

Table 2. Contrasting characteristics of the selected commercial biocontrol agents.

BCAs, Commercial Products	Mycostop	Xedavir	Polyversum
Strain	*Streptomyces griseoviridis* strain K61	*Trichoderma asperellum* strain TV1	*Pythium oligandrum* strain M1/ATCC 38472
Family	Actinomycete	Ascomycete	Pythiaceae
Kingdom	Bacteria	Fungi	Oomycete
Plant/soil of isolation	*Sphagnum* peat	Root of tomato plant	Sugar beet
Locations of isolation	Finland	Italy	Czechoslovakia
Recommended Use	General soil treatment	General soil treatment	Barley and wheat aerial treatment
Target(s)	Soil-borne pathogen	Soil-borne pathogen	*Fusarium graminearum*
Commercialization company	Lallemand Plant Care®	Xeda International®	DeSangosse®
Date of marketing	2014	2013	2015
Identified general mechanism of action	Production of antimicrobial compound, mycoparasitism, plant defense induction	Mycoparasitism, plant defense induction	Mycoparasitism, plant defense induction
/on mycotoxigenic fungi	/none	/none	/on *F. graminearum*

5.1.3. Behaviors on Different Media

To identify the culture media best suited to the dual culture bioassay of each of the BCAs/pathogen, seven culture media candidates were chosen: PDA, Czapeck agar (CZA), Czapeck yeast agar (CYA), and oat meal agar (OMA) for their affinity with pathogens [45–48], and potato dextrose agar (PDA), International *Streptomyces* Project Medium 2 (ISP2), malt yeast agar (MYA), and corn meal agar (CMA) for their affinity with BCAs (respectively all BCAs, Myco [49,50], Xeda [13,51,52], and Poly [53–55]). The compositions of media are available in Figure A1. All isolates were inoculated at the center of Petri dishes (5 µL × 10^4 spores mL^{-1}) and the growth kinetics were measured by image analysis on ImageJ software (area measurement in cm^2/every two days for 8 days). Produced spores were evaluated by flooding the plates on the final day at the end of kinetics (after 8 days) and counted under an optic microscope with Thoma cell. Mycotoxin analyses were performed for pathogen with the method described below (Section 5.2.2).

This result was be used to select two culture media for each BCA/pathogen pair for the dual culture bioassay tests.

5.2. Antagonist Activity in vitro on Selected Culture Media

5.2.1. Dual Culture Bioassay and Growth Evaluation

On selected media (Table 1), one BCA and one pathogen were inoculated in a Petri dish, 85 mm in diameter, at the same distance from the center of plate. The pathogens and BCAs were inoculated with the spore suspension except for Poly, which was inoculated with 7-day plugs (5 mm in diameter). Each BCA was inoculated 24 h in advance (5 µL × 10^5 spores mL^{-1} or plug); then the pathogens were inoculated at the opposite side (45 mm from BCA, 5 µL × 10^4 spores mL^{-1}). Plates were incubated at 25 ± 2 °C, in the dark, for 12 days. Plates inoculated only with the pathogen were used as a control treatment. Every two days, Petri dish were photographed, and the colony area of each BCA/pathogen were measured in cm^2 by image analysis using ImageJ software (1.52a, Wayne Rasband National Institute of Health, Bethesda, MD, USA, 2018). Values were used to create growth curves. Each BCA/pathogen dual culture bioassay and control was set up in triplicate for each analyzed day and two independent repetitions of the test were done.

5.2.2. Mycotoxin Extraction and Analysis

After growth evaluation, each dual culture bioassay plate was submitted to a mycotoxin analysis. The extraction procedure described by Moreau and Levi [56] was used with some modifications (no purification steps). For both pathogens, half a Petri dish was sampled by cutting along the line formed by the two inoculation points, and finely cut and weighed. For *F. graminearum* tests or controls, 50 mL of acetonitrile/water/acetic acid (79:20:1, *v/v/v*) were added. For *F. verticillioides* tests or controls, 150 mL of water/acetic acid (99.5:0.5, *v/v/v*) were added. All samples were homogenized by mechanical agitation for 20 min. *F. graminearum* samples were preliminarily diluted 1:50 in water/acetic acid (99.5:0.5; mobile phase of analyzer) and filtered with a CA filter (0.45 µm, Carl Roth GmbH, Karlsruhe, Germany); *F. verticillioides* samples were directly filtered, and were ready for injection.

For all samples, mycotoxin detection and quantification were achieved using an Ultra High-Performance Liquid Chromatography (UHPLC, Shimadzu, Tokyo, Japan) coupled with a mass spectrometer (8040, Shimadzu, Tokyo, Japan). LC separation was performed using a Phenemenex Kinetex XB Column C18 (50 mm × 2 mm; 2.6 µm particles) at 50 °C, with an injection volume of 50 µL. Mobile phase composition was (A) 0.5% acetic acid in ultra-pure water and (B) 0.5% acetic acid in isopropanol (HPLC MS grade, Sigma, St Louis, MO, USA), and the mobile phase flow rate was 0.4 mL min^{-1}. The mass spectrometer was operated in electrospray positive (ESI+) and negative (ESI–) ionization mode, and two multiple reaction monitoring (MRM) transitions for each analyte were monitored for quantification (Q) and qualification (q) (Table 3, Table 4). All data were analyzed using LabSolution Software (v5.91/2017, Shimadzu, Tokyo, Japan,2017). Limits of detection or quantification (LOD/LOQ in ng mL^{-1}, respectively) for each mycotoxin were: DON (4/14), 15-ADON (10/35), FB$_1$ (0.03/0.1), and FB$_2$ (0.01/0.05). Mycotoxin levels were expressed in ng g^{-1} of medium or ng cm^{-2}, respectively, for global and specific mycotoxin production. Values were used to create global and specific mycotoxin production curves. Each BCA/pathogen dual culture bioassay and control mycotoxin analysis was set up in triplicate for each measurement day, and two independent repetitions of the test were done.

Table 3. MS/MS parameters for isotope-labelled internal standard (IS). MRM: multiple reaction monitoring.

IS	Polarity	MRM	EC
DON C13	–	370.3 > 59.0	35
FB$_1$ C13	+	756.3 > 356.5	–47

Table 4. MS/MS parameters for mycotoxins. Q: quantification; q: qualification.

Mycotoxin	Polarity	MRM Q	EC MRM Q	MRM Q	EC MRM Q	R^2 Calibration Curve	IS
Fumonisin B$_1$	+	722.35 > 334.5	–43	722.35 > 352.0	–44	0.9982	FB$_1$ C13
Fumonisin B$_2$	+	706.2 > 318.4	–37	706.2 > 3336.4	–44	0.9980	FB$_1$ C13
DON	–	355.0 > 59.0	35	355.0 > 265.1	35	0.9998	DON C13

5.3. Macroscopic and Microscopic Observations

For the macroscopic observations, plates obtained in the same conditions described previously were used to identify for signs of an inhibition zone between each BCA/pathogen dual culture bioassay or/and the activation of secondary metabolism through the differential presence of pigments compared to the control or/and the BCA overgrowth on the pathogenic colony suggesting mycoparasite activity. The BCA control plates were added, with BCA alone in same conditions.

Microscopic observations were performed with an adapted method from Reithner [57]. On glass slides, 1.5 mL of common medium (CM) for each BCA/pathogens pair (CYA for *F. graminearum* and PDA for *F. verticillioides*) were deposited, inoculated with BCA/pathogen pairs on opposite sides (4.5 cm), incubated at 25 °C on supports in Petri dishes containing wet sterile filter paper, and sealed with parafilm. Glass slides inoculated only with pathogens were used as the control. The evolution of interaction was observed daily until 24 h after contact between BCAs and pathogens (120 h), with a Zeiss PrimoStar microscope. Pictures were taken using a Zeiss Axiocam ERc5s camera (Carl Zeiss Microscopy, Thornwood, NY, USA). The obtained pictures were treated with Photoshop CC Software (19.0, Adobe, San Jose, CA, USA, 2017), using the following pipe-line: contrast/light correction, filter layer correction, selection of a color/black and white density filter (reduces yellowing of color treatment, provides contrast for the microorganism contours), complementary manual selection of microorganisms, color filling (background transparency and Gaussian blur), and directional gradient in section.

5.4. Nutrient Profiling

Phenotype microarrays plates (Biolog, Hayward, CA, USA) were used for the nutrient profiling of all microorganisms according to the manufacturer's instructions. For the carbon and nitrogen consumption profiling, PM1 and PM3 were used, respectively. Carbon sources in PM1 are in the 5–50 mM range and nitrogen sources in PM3 are in the 2–20 mM range [58]. Inocula were suspended in sterile Biolog FF inoculating fluid and adjusted to a transmission of 81 and 62% at 590 nm for Myco and other micro-organisms, respectively. Plates were incubated in darkness at 25 °C, and growth (optical density) at 750 nm after 0, 12, 18, 24, 36, 42, 48, 60, 66, 72, 84, 90, 96, 114, and 138 h using an Enspire Multimode Reader (Perkin Elmer, Waltham, MA, USA) was monitored. The precise composition of all dual culture bioassay media was established in light of literature and confronted with the nutrient resources of plates.

5.5. Data Expression and Statistical Analysis

With the aim of considering complete growth and/or mycotoxin production kinetics, and not only a final point, the area under the growth/mycotoxin production curves were calculated for the antagonistic test and nutrient profiling test.

Statistical data analysis was performed with R Software (3.4.4, R Foundation for Statistical Computing, Vienna, Austria, 2017). Normality and homogeneity of variances were checked with the Shapiro–Wilk test (with Holm–Bonferroni correction) and Levene's test, respectively. For each pathogen, the effect of BCA treatments was tested with a one-way ANOVA and multiple comparisons of means were done with Tukey's test ($\alpha = 0.05$). For the principal component analysis (PCA), data were standardized, and the analysis was done on the correlation matrix with the FactoMine R package. The first two components were retained in both cases. For the visualization of specific differences between microorganisms nutrient profiling, the negative control was subtracted and a heatmap was built with the ggplot package. The color scale indicates differential nutrient consumption in comparison with other micro-organisms (row comparison).

Author Contributions: L.P., C.S. and S.S.-G. designed the study. N.D. performed mycotoxin analysis via HPLC MS-MS. V.M. participated in the extraction of mycotoxins from culture media. L.P. carried out the entire range of manipulations and experimental approaches and analyzed the data. L.P., C.S. and S.S.-G. interpreted the data. L.P. wrote the original draft. A.F., C.S. and S.S.-G. carefully supervised and reviewed the paper. All authors have read and agreed to the published version of the manuscript.

Funding: This research received no external funding.

Acknowledgments: Special thanks go to Maureen PELLAN, graphic designer, for her support in the analysis and enhancement of microscopy photographs. We thank the DeSangosse® Lallemand Plant Care® and Xeda International® companies for their donations of different commercial biocontrol agents (respectively Polyversum®, Mycostop®, and Xedavir®) used in different experiments, and Antofénol for the generous loan of the plate spectrophotometer.

Conflicts of Interest: The authors declare no conflict of interest.

Appendix A.

Appendix A.1. Medium Composition

CYA - Czapek Yeast Extract Agar	g/L
Sucrose	30
Agar[1]	15
Yeast Extract[2]	5
Dipotassium hydrogen phosphate	1
Sodium nitrate	0.3
Potassium chloride	0.05
Magnesium sulphate	0.05
Potassium chloride	0.05
Trace elements solution (T.E.S.)	1 ml
TES:	*g/100ml*
Iron sulphate	*0.1*
Zinc sulphate	*0.1*
Copper sulphate	*0.05*

CZA - Czapek	g/L
Sucrose	30
Agar[1]	15
Sodium nitrate	2
Dipotassium phosphate	1
Magnesium sulphate heptahydrate	0.5
Potassium chloride	0.5
Iron sulphate heptahydrate	0.01
Trace elements solution (T.E.S.)	200 µl
TES:	*g/100ml*
Citric acid monohydrate	*5*
Zinc sulphate heptahydrate	*5*
Copper sulphate pentahydrate	*0.25*
Manganese sulphate monohydrate	*0.05*
Boric acid	*0.05*
Sodium molibdate dihydrate	*0.05*

PDA – Potatoes Dextrose Agar	g/L
Dextrose/Glucose	20
Agar[1]	15
Potatoes infusion[3]	4

MYA – Malt and Yeast Extract Agar	g/L
Agar[1]	20
Malt extract[4]	20
Yeast extract[2]	1

ISP2	g/L
Agar[1]	20
Malt extract[4]	10
Glucose	4
Yeast extract[2]	4

OMA – Oat Meal agar	g/L
Oatmeal[5]	20
Agar[2]	18
Trace elements solution (T.E.S.)	1 ml
TES:	*g/100ml*
Manganese chloride	*0.1*
Iron sulphate	*0.1*
Zinc sulphate	*0.1*

CMA – Corn Meal Agar	g/L
Agar	15
Cornmeal	50

Figure A1. Detailed composition of culture media. In all tables, principal ingredients for the preparation of media in laboratory are described. Trace element solutions (TES) were added after autoclaving the culture media. For complex ingredients, superscript numbers refer to the main components in C and N identified through the literature. [1] Agar [59,60]. Major C components: β-D- galactopyrose, α-L-galactopyranose, agarobiose, agaropectine. Major N component: uronic acid. [2] Yeast extract [60,61]. Major C component: mannitol. Major N components: alanine, lysine, aspartic acid, glutamic acid. [3] Potato infusion [62,63]. Major C component: starch (glucose + glucose), Major N components: asparagine, glutamine, aspartic acid. [4] Malt extract [60,64]. Major C components: maltose, glucose-fructose, dextrine, maltobiose. Major N components: glutamic acid, proline, alanine, leucine, phenylalanine. [5] Oatmeal [65–67]. Major C components: starch (glucose+glucose), arabinoxylane, β-glucan. Major N components: phenylalanine, tyrosine, leucine, valine, arginine, lysine, glutamic acid, glycine, alanine, serine.

Appendix A.2. Kinetics of Pathogen Growth and Mycotoxin Inhibition in Dual Culture Bioassay with BCAs (Common Media, CM)

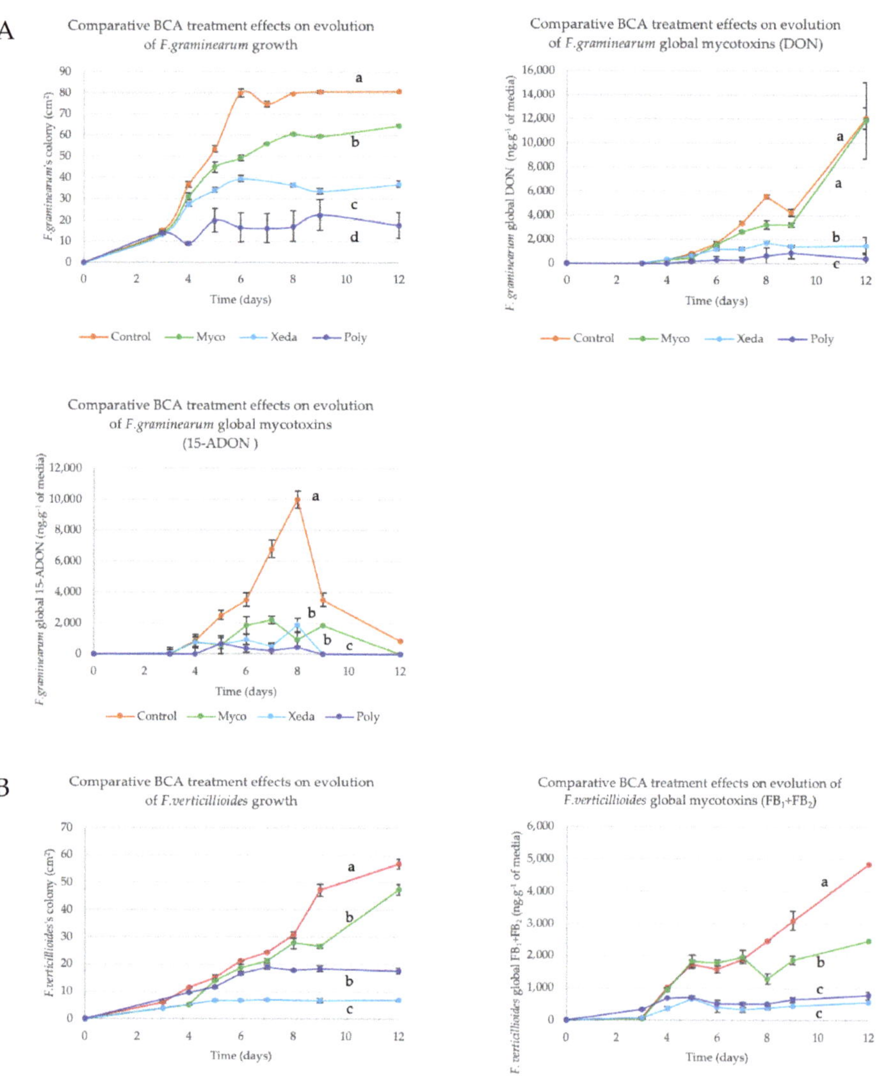

Figure A2. Comparative effects of BCA treatments on the evolution of pathogens. Dual culture bioassays for all BCAs against one pathogen were performed using common medium, CM. (**A**) *F. graminearum*, on CYA, in orange; (**B**) *F. verticillioides*, on PDA, in pink, for 12 days. Growth (A1, B1) is expressed in cm^2, and mycotoxin production levels (A2, A3, B2) are expressed in ng g^{-1} of media. FB$_1$ + FB$_2$ are represented together because the quantity of FB2 was not significant. ANOVA test, *p*-value < 0.05.

References

1. Gruber-Dorninger, C.; Jenkins, T.; Schatzmayr, G. Global Mycotoxin Occurrence in Feed: A Ten-Year Survey. *Toxins* **2019**, *11*, 375. [CrossRef] [PubMed]
2. Ferrigo, D.; Raiola, A.; Causin, R. Fusarium toxins in cereals: Occurrence, legislation, factors promoting the appearance and their management. *Molecules* **2016**, *21*, 627. [CrossRef] [PubMed]
3. Hussein, H.S.; Brasel, J.M. Toxicity, metabolism, and impact of mycotoxins on humans and animals. *Toxicology* **2001**, *167*, 101–134. [CrossRef]
4. Zain, M.E. Impact of mycotoxins on humans and animals. *J. Saudi Chem. Soc.* **2011**, *15*, 129–144. [CrossRef]
5. de Ruyck, K.; de Boevre, M.; Huybrechts, I.; de Saeger, S. Dietary mycotoxins, co-exposure, and carcinogenesis in humans: Short review. *Mutat. Res. Rev. Mutat. Res.* **2015**, *766*, 32–41. [CrossRef]
6. Gerez, J.R.; Pinton, P.; Callu, P.; Grosjean, F.; Oswald, I.P.; Bracarense, A.P.F.L. Deoxynivalenol alone or in combination with nivalenol and zearalenone induce systemic histological changes in pigs. *Exp. Toxicol. Pathol.* **2015**, *67*, 89–98. [CrossRef]
7. Sun, L.H.; Lei, M.Y.; Zhang, N.Y.; Gao, X.; Li, C.; Krumm, C.S.; Qi, D.S. Individual and combined cytotoxic effects of aflatoxin B1, zearalenone, deoxynivalenol and fumonisin B1 on BRL 3A rat liver cells. *Toxicon* **2015**, *95*, 6–12. [CrossRef]
8. Jard, G.; Liboz, T.; Mathieu, F.; Guyonvarch, A.; Lebrihi, A. Review of mycotoxin reduction in food and feed: From prevention in the field to detoxification by adsorption or transformation. *Food Addit. Contam. Part A Chem. Anal. Control Expo. Risk Assess.* **2011**, *28*, 1590–1609. [CrossRef]
9. Aktar, W.; Sengupta, D.; Chowdhury, A. Impact of pesticides use in agriculture: Their benefits and hazards. *Interdiscip. Toxicol.* **2009**, *2*, 1–12. [CrossRef]
10. Torres, A.M.; Palacios, S.A.; Yerkovich, N.; Palazzini, J.M.; Battilani, P.; Leslie, J.F.; Logrieco, A.F.; Chulze, S.N. Fusarium head blight and mycotoxins in wheat: Prevention and control strategies across the food chain. *World Mycotoxin J.* **2019**, *4*, 333–355. [CrossRef]
11. Vallet, M.; Vanbellingen, Q.P.; Fu, T.; le Caer, J.P.; Della-Negra, S.; Touboul, D.; Duncan, K.R.; Nay, B.; Brunelle, A.; Prado, S. An Integrative Approach to Decipher the Chemical Antagonism between the Competing Endophytes Paraconiothyrium variabile and Bacillus subtilis. *J. Nat. Prod.* **2017**, *80*, 2863–2873. [CrossRef] [PubMed]
12. Daguerre, Y.; Siegel, K.; Edel-Hermann, V.; Steinberg, C. Fungal proteins and genes associated with biocontrol mechanisms of soil-borne pathogens: A review. *Fungal Biol. Rev.* **2014**, *28*, 97–125. [CrossRef]
13. Matarese, F.; Sarrocco, S.; Gruber, S.; Seidl-Seiboth, V.; Vannacci, G. Biocontrol of Fusarium head blight: Interactions between Trichoderma and mycotoxigenic Fusarium. *Microbiology* **2012**, *158*, 98–106. [CrossRef] [PubMed]
14. Liu, P.; Luo, L.; Long, C. An Characterization of competition for nutrients in the biocontrol of Penicillium italicum by Kloeckera apiculata. *Biol. Control* **2013**, *67*, 157–162. [CrossRef]
15. Celar, F. Competition for ammonium and nitrate forms of nitrogen between some phytopathogenic and antagonistic soil fungi. *Biol. Control* **2003**, *28*, 19–24. [CrossRef]
16. Kim, S.H.; Vujanovic, V. Relationship between mycoparasites lifestyles and biocontrol behaviors against Fusarium spp. and mycotoxins production. *Appl. Microbiol. Biotechnol.* **2016**, *100*, 5257–5272. [CrossRef] [PubMed]
17. Shoresh, M.; Harman, G.E.; Mastouri, F. Induced Systemic Resistance and Plant Responses to Fungal Biocontrol Agents. *Annu. Rev. Phytopathol.* **2010**, *48*, 21–43. [CrossRef]
18. Lugtenberg, B.J.J.; Caradus, J.R.; Johnson, L.J. Fungal endophytes for sustainable crop production. *FEMS Microbiol. Ecol.* **2016**, *92*, 194. [CrossRef]
19. Biological Control of Mycotoxigenic Fungi and Their Toxins: An Update for the Pre-Harvest Approach. Available online: https://biblio.ugent.be/publication/8580508/file/8580795 (accessed on 26 February 2020).
20. Legrand, F.; Picot, A.; Cobo-Díaz, J.F.; Chen, W.; le Floch, G. Challenges facing the biological control strategies for the management of Fusarium Head Blight of cereals caused by F. graminearum. *Biol. Control* **2017**, *113*, 26–38. [CrossRef]
21. Picot, A.; Barreau, C.; Pinson-Gadais, L.; Caron, D.; Lannou, C.; Richard-Forget, F. Factors of the Fusarium verticillioides-maize environment modulating fumonisin production. *Crit. Rev. Microbiol.* **2010**, *36*, 221–231. [CrossRef]

22. Gardiner, D.M.; Kazan, K.; Manners, J.M. Nutrient profiling reveals potent inducers of trichothecene biosynthesis in Fusarium graminearum. *Fungal Genet. Biol.* **2009**, *46*, 604–613. [CrossRef] [PubMed]

23. Zhao, Y.; Selvaraj, J.N.; Xing, F.; Zhou, L.; Wang, Y.; Song, H.; Tan, X.; Sun, L.; Sangare, L.; Folly, Y.M.E.; et al. Antagonistic action of Bacillus subtilis strain SG6 on Fusarium graminearum. *PLoS ONE* **2014**, *9*, 1–11. [CrossRef]

24. Dal Bello, G.M.; Mónaco, C.I.; Simón, M.R. Biological control of seedling bright of wheat caused by Fusarium graminearum with beneficial rhizosphere microorganisms. *World J. Microbiol. Biotechnol.* **2002**, *18*, 627–636. [CrossRef]

25. Oskay, M. Antifungal and antibacterial compounds from Streptomyces strains. *Afr. J. Biotechnol.* **2009**, *8*, 3007–3017.

26. Igarashi, Y. Screening of Novel Bioactive Compounds from Plant-Associated Actinomycetes. *Actinomycetologica* **2004**, *18*, 63–66. [CrossRef]

27. Lu, D.; Ma, Z.; Xu, X.; Yu, X. Isolation and identification of biocontrol agent Streptomyces rimosus M527 against Fusarium oxysporum f. sp. cucumerinum. *J. Basic Microbiol.* **2016**, *56*, 929–933. [CrossRef]

28. Wang, C.; Wang, Z.; Qiao, X.; Li, Z.; Li, F.; Chen, M.; Wang, Y.; Huang, Y.; Cui, H. Antifungal activity of volatile organic compounds from Streptomyces alboflavus TD-1. *FEMS Microbiol. Lett.* **2013**, *341*, 45–51. [CrossRef]

29. Menke, J.; Weber, J.; Broz, K.; Kistler, H.C. Cellular Development Associated with Induced Mycotoxin Synthesis in the Filamentous Fungus Fusarium graminearum. *PLoS ONE* **2013**, *8*, e63077. [CrossRef]

30. Chanda, A.; Roze, L.V.; Kang, S.; Artymovich, K.A.; Hicks, G.R.; Raikhel, N.V.; Calvo, A.M.; Linz, J.E. A key role for vesicles in fungal secondary metabolism. *Proc. Natl. Acad. Sci. USA* **2009**, *106*, 19533–19538. [CrossRef]

31. Inch, S.; Gilbert, J. Scanning electron microscopy observations of the interaction between Trichoderma harzianum and perithecia of Gibberella zeae. *Mycologia* **2011**, *103*, 1–9. [CrossRef]

32. El-Komy, M.H.; Saleh, A.A.; Eranthodi, A.; Molan, Y.Y. Characterization of novel trichoderma asperellum isolates to select effective biocontrol agents against tomato fusarium wilt. *Plant Pathol. J.* **2015**, *31*, 50–60. [CrossRef] [PubMed]

33. Gajera, H.; Domadiya, R.; Patel, S.; Kapopara, M.; Golakiya, B. Molecular mechanism of Trichoderma as bio-control agents against phytopathogen system—A review. *Curr. Res. Microbiol. Biotechnol.* **2013**, *1*, 133–142.

34. Druzhinina, I.S.; Seidl-Seiboth, V.; Herrera-Estrella, A.; Horwitz, B.A.; Kenerley, C.M.; Monte, E.; Mukherjee, P.K.; Zeilinger, S.; Grigoriev, I.V.; Kubicek, C.P. Trichoderma: The genomics of opportunistic success. *Nat. Rev. Microbiol.* **2011**, *9*, 749–759. [CrossRef] [PubMed]

35. Kubicek, C.P.; Herrera-Estrella, A.; Seidl-Seiboth, V.; Martinez, D.A.; Druzhinina, I.S.; Thon, M.; Zeilinger, S.; Casas-Flores, S.; Horwitz, B.A.; Mukherjee, P.K.; et al. Comparative genome sequence analysis underscores mycoparasitism as the ancestral life style of Trichoderma. *Genome Biol.* **2011**, *12*, R40. [CrossRef] [PubMed]

36. Mukherjee, P.K. Genomics of Biological Control-Whole Genome Sequencing of Two Mycoparasitic *Trichoderma* spp. *Curr. Sci.* **2011**, *101*, 268.

37. Tian, Y.; Tan, Y.; Liu, N.; Liao, Y.; Sun, C.; Wang, S.; Wu, A. Functional agents to biologically control Deoxynivalenol contamination in cereal grains. *Front. Microbiol.* **2016**, *7*, 1–8. [CrossRef] [PubMed]

38. Wei, Z.; Yang, T.; Friman, V.P.; Xu, Y.; Shen, Q.; Jousset, A. Trophic network architecture of root-associated bacterial communities determines pathogen invasion and plant health. *Nat. Commun.* **2015**, *6*, 1–9. [CrossRef]

39. le Floch, G.; Vallance, J.; Benhamou, N.; Rey, P. Combining the oomycete Pythium oligandrum with two other antagonistic fungi: Root relationships and tomato grey mold biocontrol. *Biol. Control* **2009**, *50*, 288–298. [CrossRef]

40. Cliquet, S.; Tirilly, Y. Development of a defined medium for Pythium oligandrum oospore production. *Biocontrol Sci. Technol.* **2002**, *12*, 455–467. [CrossRef]

41. Hendrix, J.W. Sterol induction of reproduction and stimulation of growth of Pythium and Phytophthora. *Science* **1964**, *144*, 1028–1029. [CrossRef]

42. Ko, W.-H. Sexual reproduction in pythiaceous fungi Chemical stimulation of sexual reproduction in Phytophthora and Pythium. *Bot. Bull. Acadamia Singap.* **1998**, *39*, 81–86.

43. Faure, C.; Rey, T.; Gaulin, E.; Dumas, B. Pythium oligandrum: A necrotrophic mycoparasite of Fusarium graminearum. In Proceedings of the Natural Products and Biocontrol, Perpignan, France, 25–28 September 2018.

44. Crippin, T.; Renaud, J.B.; Sumarah, M.W.; Miller, J.D. Comparing genotype and chemotype of Fusarium graminearum from cereals in Ontario, Canada. *PLoS ONE* **2019**, *14*, e0216735. [CrossRef] [PubMed]

45. Leslie, J.F.; Summerell, B.A. *Fusarium Laboratory Manual*; Blackwell: Hoboken, NJ, USA, 2008; ISBN 9780813819198.

46. Lígia Martins, M.; Marina Martins, H. Influence of water activity, temperature and incubation time on the simultaneous production of deoxynivalenol and zearalenone in corn (Zea mays) by Fusarium graminearum. *Food Chem.* **2002**, *79*, 315–318. [CrossRef]

47. Droce, A.; Sørensen, J.L.; Sondergaard, T.E.; Rasmussen, J.J.; Lysøe, E.; Giese, H. PTR2 peptide transporters in Fusarium graminearum influence secondary metabolite production and sexual development. *Fungal Biol.* **2017**, *121*, 515–527. [CrossRef] [PubMed]

48. Facchini, F.D.A.; Vici, A.C.; Pereira, M.G.; Jorge, J.A.; de Moraes, M.D.L.T. Enhanced lipase production of Fusarium verticillioides by using response surface methodology and wastewater pretreatment application. *J. Biochem. Technol.* **2015**, *6*, 996–1002.

49. Shepherd, M.D.; Kharel, M.K.; Bosserman, M.A.; Rohr, J. Laboratory Maintenance of *Streptomyces* Species. *Curr. Protoc. Microbiol.* **2010**, *18*, 1–8. [CrossRef]

50. Boudjeko, T.; Tchinda, R.A.M.; Zitouni, M.; Nana, J.A.V.T.; Lerat, S.; Beaulieu, C. Streptomyces cameroonensis sp. nov., a Geldanamycin Producer That Promotes Theobroma cacao Growth. *Microbes Environ.* **2017**, *32*, 24–31. [CrossRef]

51. Wu, Q.; Sun, R.; Ni, M.; Yu, J.; Li, Y.; Yu, C.; Dou, K.; Ren, J.; Chen, J. Identification of a novel fungus, Trichoderma asperellum GDFS1009, and comprehensive evaluation of its biocontrol efficacy. *PLoS ONE* **2017**, *12*, e0179957. [CrossRef]

52. Chutrakul, C.; Alcocer, M.; Bailey, K.; Peberdy, J.F. The Production and Characterisation of Trichotoxin Peptaibols, by Trichoderma asperellum. *Chem. Biodivers.* **2008**, *5*, 1694–1706. [CrossRef] [PubMed]

53. Gerbore, J.; Vallance, J.; Yacoub, A.; Delmotte, F.; Grizard, D.; Regnault-Roger, C.; Rey, P. Characterization of Pythium oligandrum populations that colonize the rhizosphere of vines from the Bordeaux region. *FEMS Microbiol. Ecol.* **2014**, *90*, 153–167. [CrossRef] [PubMed]

54. El-Katatny, M.; Abdelzaher, H.; Shoulkamy, M. Antagonistic actions of Pythium oligandrum and Trichoderma harzianum against phytopathogenic fungi (Fusarium oxysporum and Pythium ultimum var. ultimum). *Arch. Phytopathol. Plant Prot.* **2006**, *39*, 289–301. [CrossRef]

55. Horner, N.R. Molecular Studies of the Oomycete Biocontrol Agent *Pythium oligandrum*. Ph.D. Thesis, University of Aberdeen, Aberdeen, UK, 2007.

56. Moreau, S.; Levi, M. *Highly Sensitive and Rapid Simultaneous Method for 45 Mycotoxins in Baby Food Samples by Hplc-Ms/Ms Using Fast Polarity Switching (PO-CON1480E)*; Shimadzu: Tokyo, Japan, 2014.

57. Reithner, B.; Schuhmacher, R.; Stoppacher, N.; Pucher, M.; Brunner, K.; Zeilinger, S. Signaling via the Trichoderma atroviride mitogen-activated protein kinase Tmk1 differentially affects mycoparasitism and plant protection. *Fungal Genet. Biol.* **2007**, *44*, 1123–1133. [CrossRef] [PubMed]

58. Bochner, B.R.; Gadzinski, P.; Panomitros, E. Phenotype Microarrays for high-throughput phenotypic testing and assay of gene function. *Genome Res.* **2001**, *11*, 1246–1255. [CrossRef] [PubMed]

59. Lahaye, M.; Rochas, C. Chemical structure and physico-chemical properties of agar. In *International Workshop on Gelidium*; Springer: Dordrecht, The Netherlands, 1991; pp. 137–148.

60. Bridson, E.Y.; Brecker, A. Design and Formulation of Microbial Culture Media. *Methods Microbiol.* **1970**, *3*, 229–295.

61. Filtenborg, O.; Frisvad, J.C.; Thrane, U. The Significance of Yeast Extract Composition on Metabolite Production in Penicillium. *Mod. Concepts Penicillium Aspergillus Classif.* **1990**, *185*, 433–441.

62. Tester, R.F.; Karkalas, J.; Qi, X. Starch—Composition, fine structure and architecture. *J. Cereal Sci.* **2004**, *39*, 151–165. [CrossRef]

63. Zhu, F.; Cai, Y.-Z.; Ke, J.; Corke, H. Compositions of phenolic compounds, amino acids and reducing sugars in commercial potato varieties and their effects on acrylamide formation. *J. Sci. Food Agric.* **2010**, *90*, 2254–2262. [CrossRef]

64. Robbins, G.S.; Pomeranz, Y. Amino Acid Composition of Malted Cereals and Malt Sprouts. *Proc. Annu. Meet. Am. Soc. Brew. Chem.* **1971**, *29*, 15–21. [CrossRef]

65. Sterna, V.; Zute, S.; Brunava, L. Oat Grain Composition and its Nutrition Benefice. *Agric. Agric. Sci. Procedia* **2016**, *8*, 252–256. [CrossRef]

66. Chwil, S. A study on the effects of foliar feeding under different soil fertilization conditions on the yield structure and quality of common oat (Avena sativa L.). *Acta Agrobot.* **2014**, *67*, 109–120. [CrossRef]
67. Lapvetelainen, A.; Aro, T. Protein composition and functionality of high-protein oat flour derived from integrated starch-ethanol process. *Cereal Chem.* **1994**, *71*, 133–138.

Article

In-Vitro Application of a Qatari *Burkholderia cepacia* strain (QBC03) in the Biocontrol of Mycotoxigenic Fungi and in the Reduction of Ochratoxin A biosynthesis by *Aspergillus carbonarius*

Randa Zeidan [1], Zahoor Ul-Hassan [1], Roda Al-Thani [1], Quirico Migheli [2] and Samir Jaoua [1,*]

[1] Department of Biological & Environmental Sciences, College of Arts & Sciences, Qatar University, Doha P.O. Box 2713, Qatar; rz1604991@student.qu.edu.qa (R.Z.); zahoor@qu.edu.qa (Z.U.-H.); ralthani@qu.edu.qa (R.A.-T.)

[2] Dipartimento di Agraria, Università degli Studi di Sassari, 07100 Sassari, Italy; qmigheli@uniss.it

* Correspondence: samirjaoua@qu.edu.qa; Tel.: +974-44034536

Received: 19 October 2019; Accepted: 26 November 2019; Published: 2 December 2019

Abstract: Mycotoxins are secondary metabolites produced by certain filamentous fungi, causing human and animal health issues upon the ingestion of contaminated food and feed. Among the safest approaches to the control of mycotoxigenic fungi and mycotoxin detoxification is the application of microbial biocontrol agents. *Burkholderia cepacia* is known for producing metabolites active against a broad number of pathogenic fungi. In this study, the antifungal potential of a Qatari strain of *Burkholderia cepacia* (QBC03) was explored. QBC03 exhibited antifungal activity against a wide range of mycotoxigenic, as well as phytopathogenic, fungal genera and species. The QBC03 culture supernatant significantly inhibited the growth of *Aspergillus carbonarius*, *Fusarium culmorum* and *Penicillium verrucosum* in PDA medium, as well as *A. carbonarius* and *P. verrucosum* biomass in PDB medium. The QBC03 culture supernatant was found to dramatically reduce the synthesis of ochratoxin A (OTA) by *A. carbonarius*, in addition to inducing mycelia malformation. The antifungal activity of QBC03's culture extract was retained following thermal treatment at 100 °C for 30 min. The findings of the present study advocate that QBC03 is a suitable biocontrol agent against toxigenic fungi, due to the inhibitory activity of its thermostable metabolites.

Keywords: Ochratoxin A; biological control; Qatari microflora; *Burkholderia cepacia*; thermostability

Key Contribution: The Qatari *Burkholderia cepacia* QBC03 has an antagonistic activity against a broad range of fungal genera and species. Moreover, the antifungal compounds produced by QBC03 are thermostable and present inhibitory activity towards both fungal growth and mycotoxin biosynthesis.

1. Introduction

Mycotoxins are natural contaminants produced by certain filamentous fungi, mainly belonging to the genera *Aspergillus*, *Penicillium*, and *Fusarium*. The contamination of food commodities with toxigenic fungal species, either at pre- or post-harvest (transportation, storage) phases, leads to the accumulation of their toxic secondary metabolites [1,2]. Among a long list of mycotoxins, aflatoxins (AFs), ochratoxin A (OTA), deoxynivalenol (DON), trichothecene (T2) and 3'-Hydroxy T2 (HT2) citrinin, patulin, fumonisins, zearalenone, and trichothecenes are widely studied [3]. The health implications due to dietary exposure to mycotoxins include immunosuppression, carcinogenicity, mutagenicity, teratogenicity, genotoxicity, etc. [3–6].

OTA, the most toxic among the ochratoxins, is a secondary metabolite of some *Aspergillus* (*A. carbonarius*, *A. ochraceus*, *A. westerdijkiae* and *A. niger*) and *Penicillium* (*P. verrucosum*, *P. nordicum*) species [7]. The contamination of these fungi leads to the accumulation of mycotoxin in a variety of food and feed products including cereals, fruits juices, animal feed, wine, and baby food [8–10]. OTA is primarily known for its nephrotoxic activity, while the other effects range from the mild reduction in animal production performance to carcinogenesis [11]. In most developed, and in some developing countries, strict regulatory measures are in practice to monitor and control OTA in food and feed. The International Agency for Research on Cancer (IARC) has placed OTA in group 2B of the list of possible human carcinogens [12].

Apart from the toxigenic potential of the fungi, several other environmental factors—such as nutrient availability, humidity, temperature, and the pH of the substrate—play a vital role in the accumulation of toxins [13–15]. Several approaches, including the use of chemicals (pesticides and fungicides), physical interventions and improved managemental technologies are in practice to reduce/eliminate the fungal disease burden [16]. Each approach holds its success rate, but the side effects on the environment and production system have been compromised to some degree. It is known that the intensive use of chemicals not only raises human health concerns due to their residual transfer to food and feed, but it also leads to the emergence of fungicide-resistant populations. In recent years, research efforts were focused to find out safe approaches to overcome the problems related to mycotoxigenic fungi. Many microbial antifungal environment-friendly products (both volatiles and non-volatiles) are being explored and tested for their potential application in agriculture and the food industry [17–20]. The mode of action of these biological agents is either antibiosis (through the production of antibiotics, lytic enzymes, and antagonistic proteins), competition for space and nutrients, and/or the enhancement of plant defense mechanisms [21]. Several bacterial biocontrol agents including *Lactobacillus*, *Pseudomonas*, and *Bacillus* spp. have been explored for the ability to produce antifungal compounds that can lyse either the fungal cell wall or cell membrane [22].

In the present study, we aimed to explore the antifungal potential of *Burkholderia cepacia* strain QBC03, isolated from the Qatari feed market, against key mycotoxigenic and plant pathogenic fungi belonging to genera *Aspergillus*, *Penicillium* and *Fusarium*.

2. Results

2.1. Determination of the Spectrum of Burkholderia Cepacia QBC03 Antifungal Activity

The antagonistic spectrum of QBC03 was explored against 21 fungal species belonging to the *Aspergillus*, *Fusarium* and *Penicillium* genera, using a spore overlay method. The bacterial compound diffused into the medium showed a clear zone of fungal growth inhibition against a wide range of fungi, as shown in Figure 1. Among the tested fungal genera, *Aspergillus* and *Penicillium* showed higher sensitivity compared to *Fusarium*. Based on the zone of fungal growth inhibition, *A. carbonarius* was the most sensitive species. Similarly, *P. camemberti* showed significantly higher sensitivity as compared to other tested *Penicillium* spp. As can be seen from Figure 1, *F. verticillioides* was the least sensitive species.

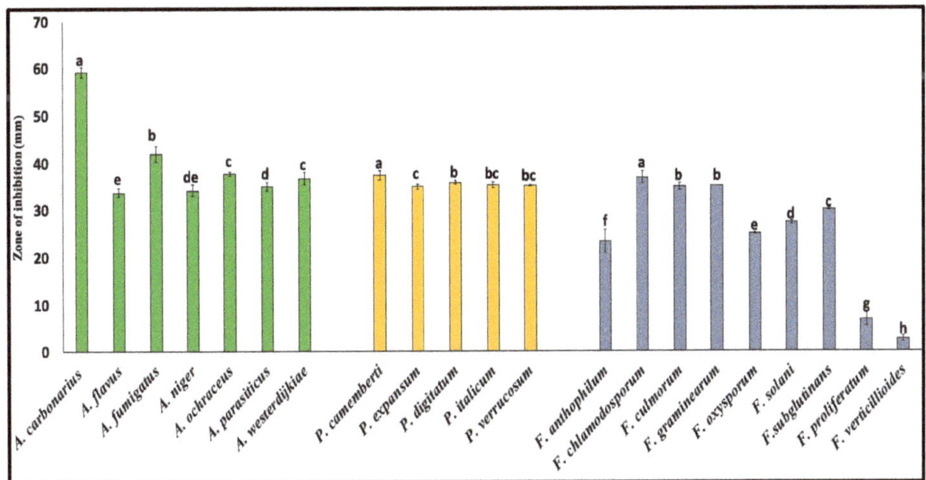

Figure 1. Effect of QBC03's antifungal compounds on fungal growth. The zones of fungal growth inhibition (mm) around the bacterial colonies were recorded at day 4 of co-incubation. *A. carbonarius* showed highest sensitivity, while *F. verticillioides* was the least sensitive fungus. The Tukey test was performed to compare the inhibition zones for species within the same genus. Bars show the zone of inhibition (mm) with denoted letters (a–h). Values sharing the same letter are non-significantly different from each other ($p \leq 0.05$).

2.2. Antifungal Activity of QBC03's Culture Extract in Solid Media

To further explore the antagonistic activity of QBC03, a bacterial-cell-free culture extract was added to PDA at different concentrations, ranging from 2.5% to 15.5%. One fungal species from each tested genus was point-inoculated on solid media. A gradual inhibition of the fungal radial growth was noticed at increasing concentrations of the bacterial culture extract. There was complete inhibition of *P. verrucosum*, *A. carbonarius* and *F. culmorum* spore germination on PDA containing bacterial culture extract at concentrations higher than 5.5%, 7.5% and 13.5%, respectively (Figure 2). The PDA plates with QBC extracts above the threshold levels inhibited fungal spore germination, even after one month of incubation, hence indicating the long-term stability of the antagonistic compounds.

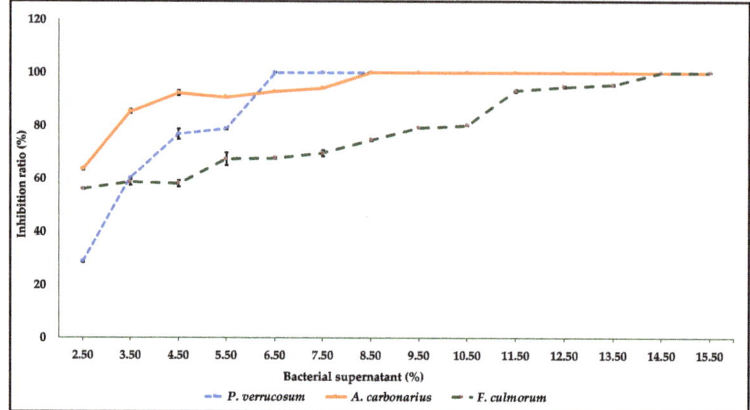

Figure 2. Comparison of the growth inhibition ratios for *A. carbonarius*, *F. culmorum* and *P. verrucosum* on PDA amended with the QBC03 culture supernatant. The PDA was amended with 2.5%, 3.5%, 4.5%, 5.5%, 6.5%, 7.5%, 8.5%, 9.5%, 9.5%, 10.5%, 11.5%, 12.5%, 13.5%, 14.5%, and 15.5% of the QBC03 culture supernatant, and the fungal colony diameter (mm) was measured after 5 days of incubation at 25 °C. The fungal colony size on treated media was compared with that on untreated PDA media to calculate the ratio of inhibtition (%).

2.3. Effect of QBC03's Metabolites on the Fungal Biomass

To analyze the antifungal activity of QBC03 in a liquid medium, fungal spores of *A. carbonarius* and *P. verrucosum* were inoculated in PDB amended with increasing concentrations of the bacterial-cell-free culture supernatant. The fungal biomass, mycelial morphology and OTA synthesis (by *A. carbonarius* only) were recorded after 72 h of incubation with shaking (approximately 0.55 xg). The addition of 1% and 2% of bacterial extract showed a drastic effect on the biomasses of *A. carbonarius* and *P. verrucosum*, respectively, which decreased to half the biomass of the control. The biomass decrease displayed a dose-dependent manner (Figure 3A). Likewise, the effect of increasing bacterial concentrations was also found to affect the synthesis of OTA by *A. carbonarius*. As shown in Figure 3B, the concentration of OTA synthesized in the liquid media by *A. carbonarius* was directly related to its biomass. Following the fungal biomass pattern, a complete inhibition of OTA synthesis by *A. carbonarius* was noted in media containing 100% bacterial supernatant.

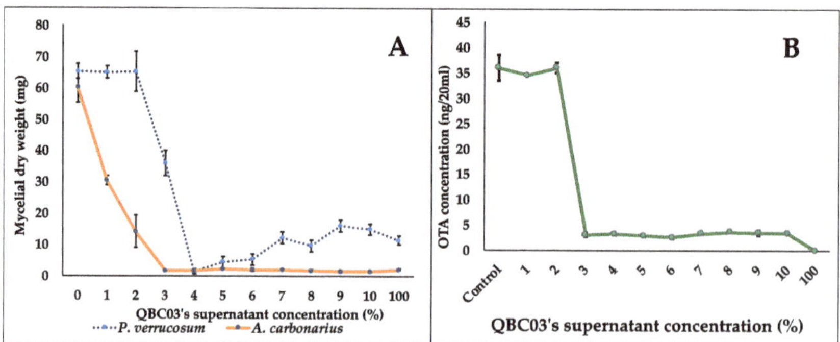

Figure 3. Effect of increasing QBC03 culture supernatant on fungal biomass and OTA synthesis in liquid media. Dry weight of the fungal mycelium (mg) in a medium inoculated with 0%, 1%, 2%, 3%, 4%, 5%, 6%, 7%, 8%, 9%, 10%, and 100% of QBC03 culture supernatant in PDB (**A**). OTA concentration (ng/20 mL PDB) synthesized by *A. carbonarius* in media containing the bacterial supernatant at levels given above (**B**).

The microscopic examination of fungal mycelia, collected from the PDB media flask amended with increasing concentrations of QBC03 culture extract, showed significant morphological alterations. The mycelia of *P. verrucosum* obtained from PDB amended with 2% QBC03 extract (Figure 4B) showed the fragmentation and shortening of most of the fungal cells with thicker walls, as compared to the untreated mycelia (Figure 4A). Similarly, Figure 4D,E show that the addition of the QBC03 extract at 2% in PDB resulted in the thickening of mycelial walls with swollen ends and bulbous protoplasmic aggregations in *A. carbonarius*, compared to the thin long mycelia of the control fungi (Figure 4C).

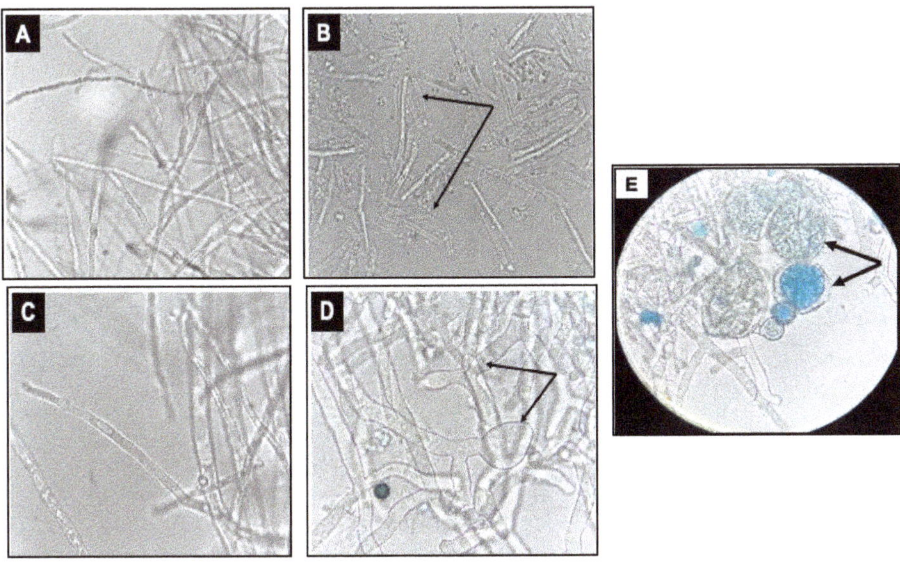

Figure 4. Effect of QBC03 culture supernatant on fungal mycelium morphology. Microscopic appearance of untreated *P. verrucosum* mycelium with intact long and thin cells (**A**), as compared to the fragmented mycelia ((**B**), arrows) of fungi treated with 2% of the bacterial extract. The morphology of *A. carbonarius* mycelia with long and thin cells from control (**C**) vs treated with 2% QBC03 extract ((**D**), arrows and (**E**), arrows). The images were captured at 1000× using a light microscope.

2.4. Effect of QBC03 Culture Extract on Fungal Spore Germination

The effect of QBC03 metabolites on the spore germination of the selected toxigenic fungal species—*P. verrucosum*, *A. westerdijikae*, *A. carbonarius* and *F. oxysporum*—was explored. In the wells of a 24-well plate, fungal spores were suspended in PDB mixed with the bacterial culture supernatant. The conidial germination was observed under a microscope and the representative images are presented in Figure 4. After 24 h of incubation, the conidia of the control groups (Figure 5A,B) showed long, protruding germination tubes, which were not found in the treated conidia (Figure 5C,D).

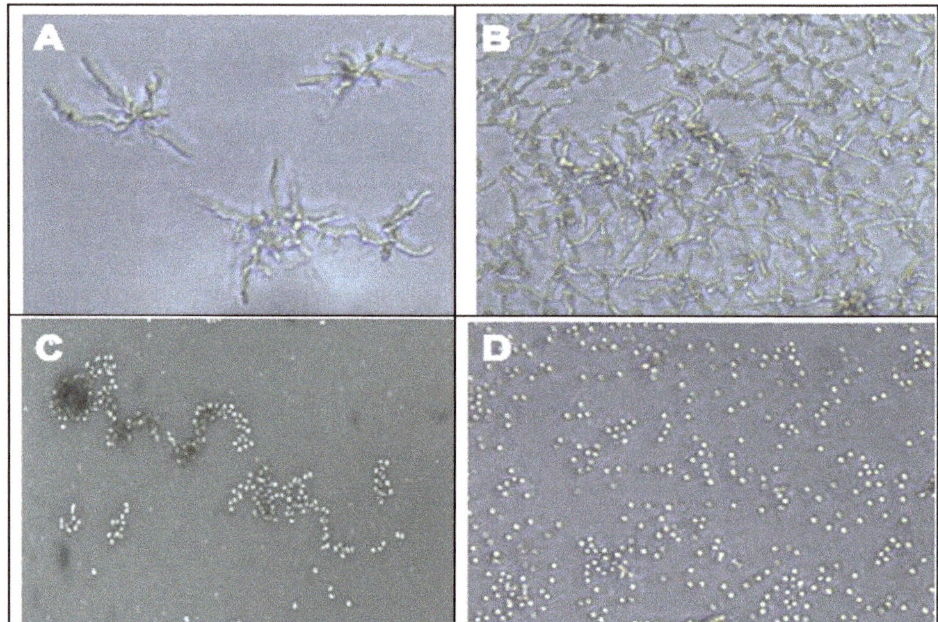

Figure 5. Effect of QBC03's supernatant on the conidiospore germination. *P. verrucosum* control spores showing germination tubes (**A**), as compared to those exposed to bacterial antifungal compounds (**C**) showing no germination after 24 hr of incubation. Plate (**B,D**) are *A. westerdijikae* spores in PDB with visible spore germination, and in PDB+QBC03 culture supernatant without any germination, respectively. The images were captured using an inverted microscope (×600).

2.5. Study on the Thermostability of QBC03 Antifungal Compounds

The thermostability of QBC03 antifungal compounds was studied by exposing its cell-free culture supernatant to different temperatures. Low-temperature treatments, such as −80 °C, −20 °C, and 4 °C for 30 min, had no significant effect on the antifungal activity of the bacterial supernatant (Figure 6A). The QBC03 culture supernatant, treated at 26 °C for 30 min, showed the highest activity against *A. carbonarius* (Figure 6A,B), and the one treated at 30 °C showed the most prominent activity towards *P. verrucosum* and *F. culmorum*. Increasing the heat treatments, at 40 °C and above, showed a decline in the antifungal efficacy of the bacterial compounds. Thermal treatment at 100 °C for 30 min, although significantly reducing the antifungal activity of the bacterial culture extract, still resulted in significant fungal growth inhibition.

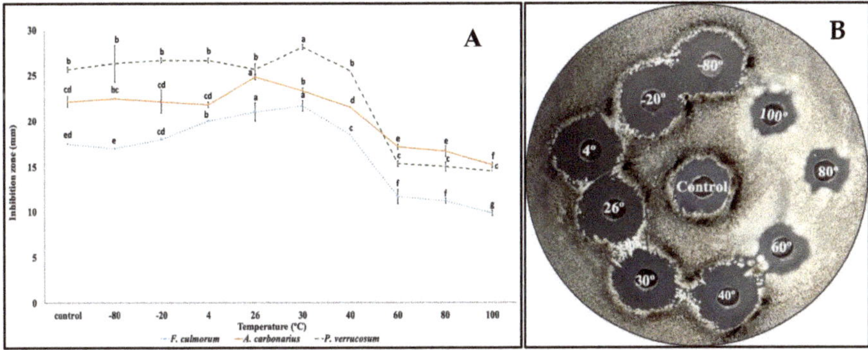

Figure 6. Effect of temperature on the stability of the QBC03 supernatant. The zones of fungal growth inhibition (mm) against *F. culmorum*, *A. carbonarius* and *P. verrucosum* were recorded around the wells containing heat-treated bacterial culture extract. (**A**) Inhibition zones at different temperature treatments for each species were denoted with letters (a–g), and inhibition zones sharing the same letter within the same species are considered as insignificantly different from each other. (**B**) Representative Petri dish of *A. carbonarius* showing the inhibition zones.

3. Discussion

In this study, we aimed to explore the antifungal potential of a *B. cepacia* strain (QBC03) isolatedfrom Qatar against some key mycotoxigenic and plant pathogenic fungi. In preliminary screening experiments, the antagonistic activity of QBC03 was tested on a total of 21 fungal species belonging to the genera *Aspergillus*, *Fusarium* and *Penicillium*. Although QBC03 showed a strong antagonistic activity against all the tested fungal species, some genus- and species-specific differences were observed. *Aspergillus* and *Penicillium* spp. were more sensitive in comparison to *Fusarium* spp. Similarly, within the same genus, some species were more sensitive than others, which might be due to structural or chemical differences in the cells. In line with the present study, the volatiles produced by *Bacillus megaterium* and *Pseudomonas protegens* showed differential effects on *Aspergillus* and *Penicillium* spp. [23].

In order to further explore the nature of the antifungal compounds, the culture supernatant of QBC03 was added at increasing concentrations to a PDA medium. Increased concentrations of the bacterial culture supernatant in the medium resulted in the increased growth inhibition of the tested fungi. There was complete inhibition of *P. verrucosum*, *A. carbonarius* and *F. culmorum* on the PDA medium amended with ≥6.5%, ≥8.5% and ≥14.5% of QBC03 culture extract, respectively. These results suggest the non-volatile nature of the antagonistic compounds released by the bacteria. In line with the present study, Kilani-Feki et al., [24] reported a 90% inhibition ratio of *Rhizoctonia solani* in PDA amended with 3% culture supernatant of *B. cepacia* (CS5). Similarly, the culture extract of *Bacillus subtilis* showed similar antagonistic activity against plant pathogenic *Stenocarpella maydis* and *Stenocarpella macrospora* [25].

In another experiment, we aimed to explore the effect of the bacterial culture supernatant on fungal biomass, mycelium morphology and mycotoxin biosynthesis. The incorporation of very low percentages (1% and 3%) of bacterial supernatant reduced the biomass significantly to almost half of that observed in the untreated control of both *A. carbonarius* and *P. verrucosum*. A similar effect was observed on *Botrytis cinerea* biomass production upon the addition of a *B. cepacia* supernatant at 0.9% [26]. Accordingly, with regards to biomass reduction, a similar decline was observed in OTA content released in the culture medium, indicating the significant effect of QBC03 metabolites on fungal growth as well as their mycotoxin production. These effects may be associated with the downregulation of fungal genes responsible for growth and mycotoxin synthesis potential. In a recent study [27], the exposure to volatile yeast (2-phenylethanol) resulted in a significant reduction (25%–99.9%) in the expression of key genes involved in OTA synthesis.

Future studies will be designed to investigate the precise nature of QBC03 antifungal compounds and their mode of action. The microscopic observation of the fungal mycelium exposed to QBC03 culture supernatant showed fragmentation, granulations and swellings. In addition, in the present study, the appearance of chlamydospores (Figure 4E), which are a general indication of fungal growth suspension and dormancy [28,29], showed that the QBC03 culture supernatant created an environment not favorable for the growth and propagation of *A. carbonarius*.

The absence of a germination tube emerging from the surface of conidia indicated that bacterial compounds completely inhibited the spore germination in *P. verrucosum*, *A. westerdijikae* and *F. oxysporum*. These findings are in line with earlier studies [30,31], reporting similar spore germination inhibition by bacterial antifungal compounds. The ability of the QBC03 antagonistic compounds to maintain their antifungal activity, even after boiling for 30 min, is the most promising feature for their possible application in future. The thermostability of bacterial compounds towards their antagonism against fungal isolates has been studied [32–35].

4. Conclusions

In this study, the in-vitro biocontrol activity of a *B. cepacia* strain (QBC03) against a wide range of fungi belonging to the genera *Aspergillus*, *Penicillium* and *Fusarium* was investigated. A high sensitivity to QBC03 was observed in fungi belonging to the *Aspergillus* genus, followed by *Penicillium* and *Fusarium*. The presence of bacterial antifungal compounds in the culture supernatant confirmed the diffusible nature of molecules. The thermostability of the QBC03 compound under a wide range of temperatures warrants its wide application. The promising antagonistic activity of QBC03 against a range of fungi suggests its application in the agriculture and the food industry, to replace or to strongly reduce the application of chemical fungicides. We believe that the findings of this research represent an excellent basis for the future exploitation of these antifungal compounds, hence, more studies need to be focused on purifying these interesting antifungal compounds and investigating their nature.

5. Materials and Methods

5.1. Microbial Strains

The *Burkholderia cepacia* strain (QBC03) is a local strain isolated from marketed feed samples in Qatar (data not shown). The list of strains included; *A. carbonarius* (AC82), *A. flavus* (CECT2687), *A. fumigatus* (AF14), *A. niger* (AN8), *A. ochraceus* (CECT2948), *A. parasiticus* (AF82), *A. westerdijkiae* (AW82), *P. camemberti* (PC44), *P. expansum* (PE82), *P. digitatum* (PD43), *P. italicum* (PI48), *P. verrucous* (TF11), *F. anthophilum* (FAn01), *F. chlamydosporum* (FCh01), *F. culmorum* (FCu11), *F. graminearum* (FGr14), *F. oxysporum* (Fox9), *F. solani* (FS05), *F. subglutinans* (FSuF12), *F. proliferatum* (FP08), and *F. verticillioides* (FV04).

5.2. Screening the Antifungal Activity of QBC03

The antifungal spectrum of QBC03 was determined using the spore overlay method around the bacterial colony. Briefly, with the help of a sterile toothpick, bacterial cells were transferred to the center of a nutrient agar plate, prepared by adding 1 g of meat extract (Mikrobiologie, Darmstadt, Germany), 5 g of peptone (Acumedia, Heywood, UK), 5 g of sodium chloride, 2 g of yeast extract (Himedia, Mumbai, India) and 15 g of agar in 1L of water. Before overlying the fungal spores, bacterial plates were incubated at 30 °C for 48 h to allow sufficient time to synthesize their antagonistic compounds. From a 7 d old, pure fungal colony, a loopful of inoculum was taken and suspended in 0.9% NaCl with 1% Tween 80. The spores were counted using a hemocytometer and their concentration was adjusted to 10^6/mL. The spores were transferred to 10 mL of soft PDA (Potato Dextrose agar from Foremedium, Hunstanton, England) and assayed around the bacterial colony. The plates were incubated at 26 °C for four days, and the diameters of the fungal zones of inhibition around the bacterial colony were measured in mm. The antifungal spectrum of QBC03 was tested against single representatives of the 21 fungal species mentioned in Section 5.1.

5.3. Investigation of QBC03 Culture Extract on the Fungal Growth in Solid Media

To test the antifungal activity of the QBC03 culture supernatant, bacterial cells were cultured in Nutrient Broth Yeast (NBY) extract media as described by Kilani-Feki and Jaoua [17]. For this purpose, a 7 h old preculture of QBC03 was prepared by transferring a single colony to 10 mL NBY broth, and from this an inoculum was taken to prepare a 48 h-old bacterial culture in NBY. The bacterial broth was centrifuged at 5500× g for 20 min and the supernatant was collected. The obtained supernatant was added to molten PDA at 2.5%, 3.5%, 4.5%, 5.5%, 6.5%, 7.5%, 8.5%, 9.5%, 9.5%, 10.5%, 11.5%, 12.5%, 13.5%, 14.5%, and 15.5%. To inhibit the growth of any bacterial cell in the media, chloramphenicol at 100 μg/L was also added. From each fungal genus, one species (*A. carbonarius*, *F. culmorum*, *P. verrucosum*) was chosen for this experiment. Three microliters of fungal spore suspension (×10³) was inoculated on the center of the PDA plates which were subsequently incubated at 26 °C for 5 days. The diameter of the fungal colonies was measured at day 5, and the inhibition ratios were estimated using the following equation:

$$\text{Inhibition ratio (\%)} = \frac{\text{diamter of control} - \text{diameter of treated}}{\text{diameterof the control}} \times 100$$

5.4. Investigation of the Effect of QBC03 Culture Extract on Fungal Biomass and Mycotoxin Synthesis

The effect of the bacterial culture supernatant on the fungal biomass production and OTA synthesis of *A. carbonarius* was studied. The bacterial culture supernatant was obtained as described in Section 5.3, and was added to PDB (Potato Dextrose Broth from Foremedium Hunstanton, England) amended with 500 μg/L chloramphenicol to obtain the final concentrations of 0%, 1%, 2%, 3%, 4%, 5%, 6%, 7%, 8%, 9%, 10%, and 100% (no PDB, only bacterial culture supernatant with antibiotic). Inocula of *A. carbonarius* or *P. verrucosum*, represented by 10 μL of a fungal spore suspension (10⁶/mL), were transferred to flasks (50 mL volume) and incubated at 26 °C for 72 h with continuous shaking at approximately 0.55× g. The fungal biomass production was measured by the filtration of flask contents using nitrocellulose paper in a Buchner funnel connected to a vacuum assembly. The biomass of the treated fungi was compared with the untreated control (only PDB). Morphological changes in the fungal mycelium including hyphal cell fragmentation, cell wall thickness, cytoplasmic degranulation, and the formation of chlamydospores were observed using a light microscope. The filtrate of each treatment was collected separately in Eppendorf tubes for the analysis of mycotoxin concentration using an OTA ELISA kit (RIDASCREEN® Ochratoxin A, R-Biopharm, Germany), and the readings were taken using an ELISA plate reader installed with SkanIt software.

5.5. Evaluation of the Effect of QBC03's Antifungal Compounds on Fungal Spore Germination

The effect of the QBC03 extract on fungal spore germination was tested in 24-wells plates. Spore suspensions were prepared from 7 day-old cultures and washed twice [22]. PDB (900 μL), amended with 500 μg/L chloramphenicol and QBC03 culture supernatant (100 μL), was added in each well. Further, 2 μL of fungal spores (×10³) from either *P. verrucosum*, *A. westerdijikae* or *F. oxysporum* were added to each well. The plate was sealed with Parafilm and incubated at 26 °C for 24 h. The germination of the spores treated with the extract was compared to the germination of the spores in control (having PDB only).

5.6. Influence of Temperature on the Stability of QBC03 Antifungal Compounds

In order to test the thermal stability of the QBC03 antifungal compounds, the bacterial culture supernatant—obtained using the method described in Section 5.2—was treated at −80 °C, −20 °C, 4 °C, 26 °C, 30 °C, 40 °C, 60 °C, 80 °C, and 100 °C for 30 min. Spore suspensions from *A. carbonarius*,

P. verrucosum and *F. culmorum* were prepared (10^6/mL), and 200 μL was spread on the surface of PDA amended with 100 μg/L chloramphenicol. Wells of 7 mm in diameter were obtained using a sterile cork-borer in the PDA plates, and 100 μL of treated extract was loaded. After 72 h, the zones of fungal growth inhibition around the wells were measured.

5.7. Statistical Analysis

SPSS statistical software (Version 23, IBM, NY, USA, 2017) was used for data analysis. An analysis of variance (ANOVA) and the multiple comparisons test (Tukey-test) were performed.

Author Contributions: Supervision, S.J.; Conceptualization, S.J., R.Z., Z.U.-H., R.A.-T.; Methodology, R.Z., S.J., Z.U.-H.; R.A.-T. and Q.M.; Validation and analysis of results, R.Z., Z.U.-H., S.J. and Q.M.; Writing, review and editing, R.Z., S.J., Z.U.-H. and Q.M.; Resources provided, S.J.

Funding: Qatar National Research Fund (a member of Qatar Foundation) under National Priorities Research Program (NPRP) grant #NPRP8-392-4-003.

Acknowledgments: This publication was made possible by National Priorities Research Program (NPRP) grant #NPRP8-392-4-003 from the Qatar National Research Fund (a member of Qatar Foundation). The statements made herein are solely the responsibility of the authors. The publication of this article was funded by the Qatar National Library.

Conflicts of Interest: The authors declare no conflict of interest.

References

1. Abbas, H.K. (Ed.) *Aflatoxin and Food Safety*; Taylor and Frances Group: Philadelphia, PA, USA, 2005.
2. Rocha, M.E.B.; Freire, F.D.C.O.; Maia, F.E.F.; Guedes, M.I.F.; Rondina, D. Mycotoxins and their effects on human and animal health. *Food Control* **2014**, *36*, 159–165. [CrossRef]
3. Anfossi, L.; Giovannoli, C.; Baggiani, C. Mycotoxin detection. *Curr. Opin. Biotechnol.* **2016**, *37*, 120–126. [CrossRef] [PubMed]
4. Zhu, R.; Zhao, Z.; Wang, J.; Bai, B.; Wu, A.; Yan, L.; Song, S. A simple sample pretreatment method for multi-mycotoxin determination in eggs by liquid chromatography tandem mass spectrometry. *J. Chromatogr.* **2015**, *1417*, 1–7. [CrossRef]
5. Hameed, M.R.; Khan, M.Z.; Khan, A.; Javed, I. Ochratoxin induced pathological alterations in broiler chicks: Effect of dose and duration. *Pak. Vet. J.* **2013**, *33*, 145–149.
6. Ul-Hassan, Z.; Khan, M.Z.; Saleemi, M.K.; Khan, A.; Javed, I.; Bhatti, S.A. Toxico-pathological effects of in-ovo inoculation of ochratoxin A (OTA) in chick embryos. *Toxicol. Pathol.* **2012**, *40*, 33–39. [CrossRef]
7. Wang, Y.; Wang, L.; Liu, F.; Wang, Q.; Selvaraj, J.N.; Xing, F.; Zhao, Y.; Liu, Y. Ochratoxin A producing fungi, biosynthetic pathway and regulatory mechanisms. *Toxins* **2016**, *8*, 83. [CrossRef]
8. Hassan, Z.U.; Roda, F.A.; Migheli, Q.; Samir, J. Detection of toxigenic mycobiota and mycotoxins in cereal feed market. *Food Control* **2018**, *84*, 389–394. [CrossRef]
9. Hassan, Z.U.; Roda, F.A.; Fathy, A.A.; Saeed, A.M.; Migheli, Q.; Samir, J. Co-occurrence of mycotoxins in commercial formula milk and cereal-based baby food. *Food Addit. Contam. Part B* **2018**, *11*, 191–197. [CrossRef]
10. De Jesus, C.L.; Bartley, A.; Welch, A.Z.; Berry, J.P. High incidence and levels of ochratoxin A in wines sourced from the United States. *Toxins* **2017**, *10*, 1. [CrossRef] [PubMed]
11. Heussner, A.H.; Bingle, L.E. Comparative ochratoxin toxicity: A review of the available data. *Toxins* **2015**, *7*, 4253–4582. [CrossRef] [PubMed]
12. Ramirez, M.L.; Cendoya, E.; Nichea, M.J.; Zachetti, V.G.L.; Chulze, S.N. Impact of toxigenic fungi and mycotoxins in chickpea: A review. *Curr. Opin. Food Sci.* **2018**, *23*, 32–37. [CrossRef]
13. Milani, J.M. Ecological conditions affecting mycotoxin production in cereals: A review. *Veterinarni Medicina* **2013**, *58*, 405–411. [CrossRef]
14. Mannaa, M.; Kim, K.D. Influence of temperature and water activity on deleterious fungi and mycotoxin producing during grain storage. *Mycobiology* **2017**, *45*, 240–254. [CrossRef] [PubMed]

15. Kabak, B.; Dobson, A.D.W.; Var, I. Strategies to prevent mycotoxin contamination of food and animal feed: A review. *Crit. Rev. Food Sci. Nutr.* **2006**, *46*, 593–619. [CrossRef] [PubMed]

16. Tola, M.; Kebede, B. Occurrence, importance and control of mycotoxins: A review. *Cogent Food Agric.* **2016**, *2*. [CrossRef]

17. Mannaa, M.; Oh, J.Y.; Kim, K.D. Biocontrol activity of volatile-producing *Bacillus megaterium* and *Pseudomonas protegens* against *Aspergillus flavus* and aflatoxin production on stored rice grains. *Mycobiology* **2017**, *45*, 213–219. [CrossRef] [PubMed]

18. Mannaa, M.; Oh, J.Y.; Kim, K.D. Microbe-mediated control of *Aspergillus flavus* in stored rice grains with a focus on aflatoxin inhibition and biodegradation. *Ann. Appl. Biol.* **2017**, *171*, 376–392. [CrossRef]

19. Hassan, Z.U.; Al-Thani, R.; Alnaimi, H.; Migheli, Q.; Jaoua, S. Investigation and application of *Bacillus licheniformis* volatile compounds for the biological control of toxigenic *Aspergillus* and *Penicillium* spp. *ACS Omega* **2019**, *4*, 17186–17193. [CrossRef]

20. Zeidan, R.; Ul-Hassan, Z.; Al-Thani, R.; Balmas, V.; Jaoua, S. Application of low-fermenting yeast *Lachancea thermotolerans* for the control of toxigenic fungi *Aspergillus parasiticus*, *Penicillium verrucosum* and *Fusarium graminearum* and their mycotoxins. *Toxins* **2018**, *10*, 242. [CrossRef]

21. Abdallah, M.F.; Ameye, M.; De Saeger, S.; Audenaert, K.; Haesaert, G. Biological control of mycotoxigenic fungi and their toxins: An update for the pre-harvest approach. In *Fungi and Mycotoxins: Their Occurrence, Impact on Health and the Economy as Well as Pre- and Postharvest Management Strategies*; Berka Njobeh, P., Ed.; IntechOpen Press: London, UK, 2018; pp. 1–31. [CrossRef]

22. Nagórska, K.; Bikowski, M.; Obuchowski, M. Multicellular behaviour and production of a wide variety of toxic substances support usage of Bacillus subtilis as a powerful biocontrol agent. *Acta Biochim. Polon.* **2007**, *54*, 495–508. [CrossRef]

23. Mannaa, M.; Kim, K.D. Biocontrol activity of volatile-producing *Bacillus megaterium* and *Pseudomonas protegens* against *Aspergillus* and *Penicillium* spp. Predominant in stored rice grains: Study II. *Mycobiology* **2018**, *46*, 52–63. [CrossRef] [PubMed]

24. Kilani-Feki, O.; Culioli, G.; Ortalo-Magné, A.; Zouari, N.; Blache, Y.; Jaoua, S. Environmental *Burkholderia cepacia* strain Cs5 acting by two analogous alkyl-quinolones and a didecyl-phthalate against a broad spectrum of phytopathogens fungi. *Curr. Microbiol.* **2011**, *62*, 1490–1495. [CrossRef] [PubMed]

25. Petatán-Sagahón, I.; Anducho-Reyes, M.A.; Silva-Rojas, H.V.; Arana-Cuenca, A.; Tellez-Jurado, A.; Cárdenas-Álvarez, I.O.; Mercado-Flores, Y. Isolation of bacteria with antifungal activity against the phytopathogenic fungi *Stenocarpella maydis* and *Stenocarpella macrospora*. *Int. J. Mol. Sci.* **2011**, *12*, 5522–5537. [CrossRef] [PubMed]

26. Kilani-Feki, O.; Jaoua, S. Biological control of *Botrytis cinerea* using the antagonistic and endophytic *Burkholderia cepacia* Cs5 for vine plantlet protection. *Can. J. Microbiol.* **2011**, *57*, 896–901. [CrossRef] [PubMed]

27. Farbo, M.G.; Urgeghe, P.P.; Fiori, S.; Marcello, A.; Oggiano, S.; Balmas, V.; Ul Hassan, Z.; Jaoua, S.; Migheli, Q. Effect of yeast volatile organic compounds on ochratoxin A-producing *Aspergillus carbonarius* and *A. ochraceus*. *Int. J. Food Microbiol.* **2018**, *284*, 1–10. [CrossRef] [PubMed]

28. Li, X.; Quan, C.S.; Fan, S.D. Antifungal activity of a novel compound from *Burkholderia cepacia* against plant pathogenic fungi. *Lett. Appl. Microbiol.* **2007**, *45*, 508–514. [CrossRef]

29. Oliveira, R.R.; Aguiar, B.D.M.; Tessmann, D.J.; Pujade-Renaud, V.; Vida, J.B. Chlamydospore formation by *Corynespora cassiicola*. *Trop. Plant Pathol.* **2012**, *37*, 415–418. [CrossRef]

30. Chen, N.; Jin, M.; Qu, H.M.; Chen, Z.Q.; Chen, Z.L.; Qiu, Z.G.; Wang, X.W.; Li, J.W. Isolation and characterization of *Bacillus* sp. producing broad-spectrum antibiotics against human and plant pathogenic fungi. *J. Microbiol. Biotechnol.* **2012**, *22*, 256–563. [CrossRef]

31. Joo, H.J.; Kim, H.Y.; Kim, L.H.; Lee, S.; Ryu, J.G.; Lee, T. A *Brevibacillus* sp. antagonistic to mycotoxigenic *Fusarium* spp. *Biol. Control* **2015**, *87*, 64–70. [CrossRef]

32. Kilani-Feki, O.; Zouari, I.; Culioli, G.; Ortalo-Magné, A.; Zouari, N.; Blache, Y.; Jaoua, S. Correlation between synthesis variation of 2-alkylquinolones and the antifungal activity of a *Burkholderia cepacia* strain collection. *World J. Microbiol. Biotechnol.* **2012**, *28*, 275–281. [CrossRef]

33. Cheba, B.A.; Zaghloul, T.I.; EL-Mahdy, A.R.; EL-Massry, M.H. Effect of pH and temperature on *Bacillus* sp. R2 chitinase activity and stability. *Procedia Technol.* **2016**, *22*, 471–477. [CrossRef]

34. Muhialdin, B.J.; Hassan, Z.; Saari, N. In vitro antifungal activity of lactic acid bacteria low molecular peptides against spoilage fungi of bakery products. *Ann. Microbiol.* **2018**, *68*, 557–567. [CrossRef]
35. Gomaa, E.Z.; Abdelall, M.F.; El-Mahdy, O.M. Detoxification of aflatoxin B1 by antifungal compounds from *Lactobacillus brevis* and *Lactobacillus paracasei*, isolated from dairy products. *Probiotics Antimicro.* **2018**, *10*, 201–209. [CrossRef] [PubMed]

Article

Fengycin Produced by *Bacillus amyloliquefaciens* FZB42 Inhibits *Fusarium graminearum* Growth and Mycotoxins Biosynthesis

Alvina Hanif [1], Feng Zhang [1], Pingping Li [1], Chuchu Li [1], Yujiao Xu [1], Muhammad Zubair [1], Mengxuan Zhang [1], Dandan Jia [1], Xiaozhen Zhao [1], Jingang Liang [2], Taha Majid [1], Jingyuau Yan [1], Ayaz Farzand [1], Huijun Wu [1], Qin Gu [1,*] and Xuewen Gao [1,*]

[1] Department of Plant Pathology, College of Plant Protection, Nanjing Agricultural University, Key Laboratory of Integrated Management of Crop Diseases and Pests, Ministry of Education, Nanjing 210095, China; rao.alvina@yahoo.com (A.H.); 2017202058@njau.edu.cn (F.Z.); 2017102044@njau.edu.cn (P.L.); 2018102014@njau.edu.cn (C.L.); 2018102015@njau.edu.cn (Y.X.); Zubair_biotech@yahoo.com (M.Z.); 12115229@njau.edu.cn (M.Z.); 2017802180@njau.edu.cn (D.J.); 2016202006@njau.edu.cn (X.Z.); tahamajid1705@yahoo.com (T.M.); 12115231@njau.edu.cn (J.Y.); ayaz.farzand@uaf.edu.pk (A.F.); hjwu@njau.edu.cn (H.W.)

[2] Development Center of Science and Technology, Ministry of Agriculture and Rural Affairs, Beijing 100176, China; liangjingang@agri.gov.cn

* Correspondence: guqin@njau.edu.cn (Q.G.); gaoxw@njau.edu.cn (X.G.); Tel.: +86-025-8439-5268 (Q.G.)

Received: 25 April 2019; Accepted: 17 May 2019; Published: 24 May 2019

Abstract: *Fusarium graminearum* is a notorious pathogen that causes Fusarium head blight (FHB) in cereal crops. It produces secondary metabolites, such as deoxynivalenol, diminishing grain quality and leading to lesser crop yield. Many strategies have been developed to combat this pathogenic fungus; however, considering the lack of resistant cultivars and likelihood of environmental hazards upon using chemical pesticides, efforts have shifted toward the biocontrol of plant diseases, which is a sustainable and eco-friendly approach. Fengycin, derived from *Bacillus amyloliquefaciens* FZB42, was purified from the crude extract by HPLC and further analyzed by MALDI-TOF-MS. Its application resulted in structural deformations in fungal hyphae, as observed via scanning electron microscopy. In planta experiment revealed the ability of fengycin to suppress *F. graminearum* growth and highlighted its capacity to combat disease incidence. Fengycin significantly suppressed *F. graminearum*, and also reduced the deoxynivalenol (DON), 3-acetyldeoxynivalenol (3-ADON), 15-acetyldeoxynivalenol (15-ADON), and zearalenone (ZEN) production in infected grains. To conclude, we report that fengycin produced by *B. amyloliquefaciens* FZB42 has potential as a biocontrol agent against *F. graminearum* and can also inhibit the mycotoxins produced by this fungus.

Keywords: fungal-bacterial interactions; *Bacillus amyloliquefaciens*; *Fusarium graminearum*; Fengycin; mycotoxins

Key Contribution: Fengycin produced from *Bacillus amyloliquefaciens* FZB42 can inhibit *F. graminearum* growth and pathogenicity and have negative impact on mycotoxins biosynthesis.

---

1. Introduction

China is the leading producer of wheat worldwide, and *Fusarium graminearum* is a major causal agent of Fusarium head blight (FHB) epidemics in the country, affecting various cereal crops, either in the field or upon their storage in humid conditions [1,2]. Infections in the field can occur at any stage, from anthesis to kernel development, and this plant pathogenic fungus mainly infects florets. Under favorable environmental conditions, an infection can be established within 3 to 4 days. *F. graminearum* produces numerous potentially important mycotoxins. Deoxynivalenol (DON) is the most abundant form of the trichothecenes found in grain and is a sesquiterpenoid. Acetylated derivatives of DON, less toxic than DON, are also found in grains 15-ADON. While trichothecenes are known to be produced during the early stages of the infection process in host plants, the most common non-steroidal estrogenic mycotoxins and Zearalenol (ZEN) is produced at the end of the infection process [3]. Fungus invades and colonizes grains and produces deoxynivalenol [4], which is the most common mycotoxin. Infected seeds show reduced germination and produce weaker seedlings. DON is the final product of the trichothecene biosynthetic pathway. It causes several biological disturbances and acts as an inhibitor during protein synthesis [5]; moreover, it is highly toxic, and thus unfit for the consumption of humans or animals [1,6]. FHB management remains challenging. There are still very few varieties of wheat that are highly resistant to *F. graminearum*. Synthetic chemicals are effective for controlling FHB in wheat; however, they are inevitably associated with environmental pollution and resistance development in *F. graminearum* [7]. Therefore, to control FHB in wheat, it is of high urgency to explore alternative management strategies that are not only reliable but also less toxic to the environment.

To date, biocontrol agents have attracted huge scientific attention as they are environmentally friendly [8]. Plant growth-promoting rhizobacteria (PGPR) are evidently promising for suppressing various fungal diseases and stimulating plant growth. Many PGPR strains have been successfully formulated as biopesticides to control plant diseases [9]. *Bacillus* spp. are the most promising antagonistic PGPR. *Bacillus amyloliquefaciens* FZB42 (now called *B. amyloliquefaciens* subsp. *plantarum* FZB42) is a Gram-positive strain and well known for its antagonistic activity, extensive rhizosphere colonization, and plant growth stimulation [10,11]. This strain reportedly produces secondary metabolites that suppress soil-borne plant pathogens; genome analysis of FZB42 revealed 10 gene clusters, covering nearly 10% of the whole genome, and these are responsible for producing secondary metabolites that display antimicrobial and nematocidal activities. These secondary metabolites include three lipopeptides (surfactin, bacillomycin D, and fengycin), three polyketides (macrolactin, bacillaene, and difficidin) [3,4], one siderophore (bacillibactin), one antibacterial dipeptide (bacilysin), and two ribosomally produced and post-translationally modified peptides plantazolicin and amylocyclicin. FZB42 can also synthesize plant hormones, such as indole-3-acetic acid, and produce volatile compounds, such as 2,3-butanediol, to promote plant growth. All these metabolites contribute to the biocontrol properties of FZB42. Moreover, this strain displays strong antagonistic activity against fungi, such as *Rhizoctonia solani*, *Botrytis cinereal* [6], *F. oxysporum* [7], and against bacteria, such as *Erwinia amylovora* [8] and *Xanthomonas oryzae* [12]. A recent study demonstrated that bacillomycin D is involved in antagonistic interactions with *F. graminearum*, provoking physiological and metabolic changes during the antagonism [13]. The living spores of FZB42 have also been used to develop commercial products, such as RhizoVital®. Accordingly, FZB42 seems to be a good candidate for use as a biocontrol agent against plant pathogens in agricultural production systems.

Here we report that fengycin produced by *B. amyloliquefaciens* FZB42 significantly inhibits the growth of *F. graminearum* and the biosynthesis of the mycotoxins, including deoxynivalenol (DON), 3-acetyldeoxynivalenol (3-ADON), 15-acetyldeoxynivalenol (15-ADON), and zearalenone (ZEN).

2. Results

2.1. Fengycin Produced by B. amyloliquefaciens FZB42 mutant AK1S Displayed Antagonistic Activity Against F. graminearum

It has already been reported that both *B. amyloliquefaciens* FZB42 and its crude extract of secondary metabolites could suppress *F. graminearum* growth. The mutant AK2 and AK1S cultures and their crude extracts showed inhibition activity against *F. graminearum* growth, as indicated by clear zones in the inoculated bacteria and extracts (Figure 1).

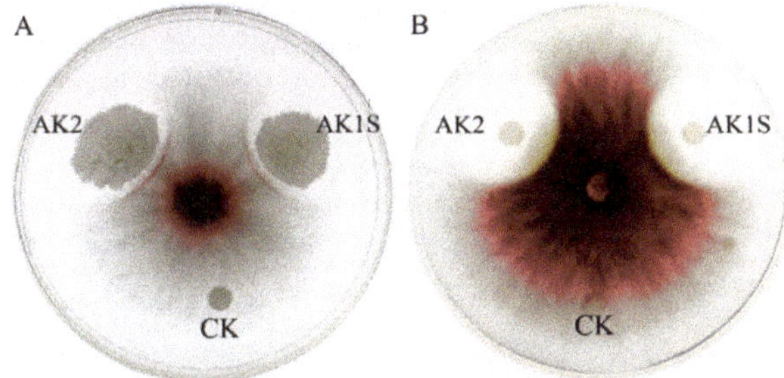

Figure 1. Antagonistic activities of AK2 and AKIS against *F. graminearum* PH-1 (**A**) and of their secondary metabolite extract (**B**). CK, control (LB medium or methanol).

AK2 could produce bacillomycin D and surfactin, but not fengycin; HPLC results indicated that AK2 showed typical peaks for bacillomycin D (from 16 to 20 min) and surfactin (from 40 to 48 min) (Figure 2). MALDI-TOF-MS analysis also confirmed that AK2 could only produce bacillomycin D and surfactin. There were peaks $(M + H)^+$ for molecular ion peaks $(M + Na)^+$ for C_{14}–C_{15} surfactin at *m/z* 1044 and 1058, and ion peaks $(M + K)^+$ for C_{15} surfactin at *m/z* 1074 (Figure 2). Furthermore, there were molecular ion peaks $(M + Na)^+$ for C_{15} bacillomycin D at *m/z* 1067, and ion peaks $(M + K)^+$ for C_{15}–C_{16} bacillomycin D at *m/z* 1083 and 1097 (Figure 2).

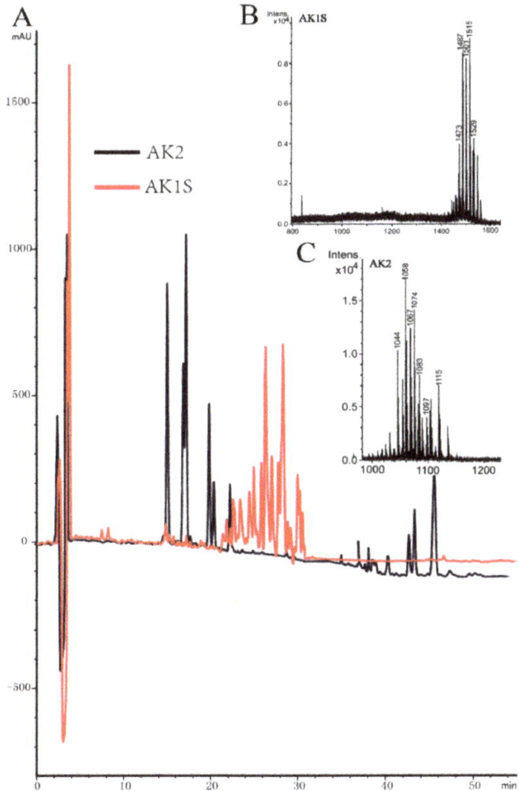

Figure 2. Analysis of lipopeptides produced by AK2 and AKIS using HPLC (**A**) and MALDI-TOF-MS (**B,C**).

Among these molecules, all containing the same ion, there was a 14-Da difference in molecular weight, suggesting the presence of varying lengths of fatty acid chains within bacillomycin D (CH2 = 14 Da). AK1S, containing double mutation, could produce fengycin, but not bacillomycin D or surfactin, as confirmed via HPLC analysis, which showed peaks only for fengycin (from 24 to 30 min). MALDI-TOF-MS confirmed this result. There were molecular ion peaks $(M+K)^+$ for Ala-6-C_{14}–C_{18} fengycin at *m/z* 1473, 1487, 1501, 1515, and 1529 (Figure 2). Our results indicated that both fengycin and bacillomycin D act as fungicidal factors and cause in vitro suppression of *F. graminearum* growth. This result coincides with the results of our previous study [13].

2.2. Ultrastructural Changes Caused by Fengycin in F. graminearum Hyphae

To elucidate the mechanism by which fengycin affects *F. graminearum* hyphal growth, we observed the morphological variations in fungal mycelia using scanning electron microscopy [14]. The micrographs showed that fengycin triggered a range of abnormalities in *F. graminearum* hyphae; fengycin-treated hyphae showed considerable deformation—they were thin and twisted, and some parts along the hyphal walls were ruptured (Figure 3). On the other hand, the micrographs of the untreated control demonstrated healthy, dense, and cylindrical hyphae.

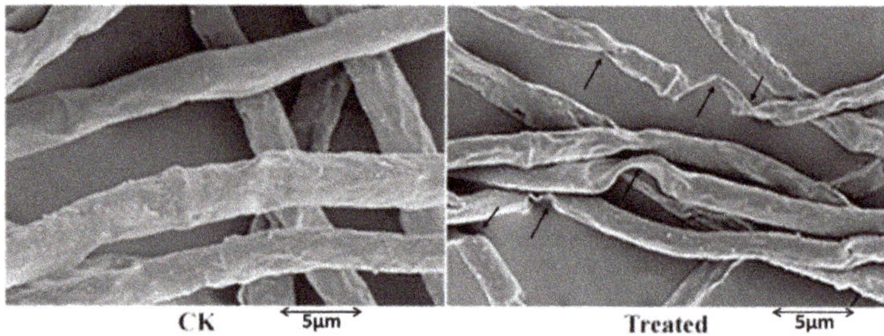

Figure 3. Effect of fengycin (90 µg/mL) on *F. graminearum* hyphae. *F. graminearum* hyphae were treated with pure fengycin, and electron micrographs were obtained. The hyphal morphology of *F. graminearum* was altered by fengycin; in comparison with the control, several deformed hyphal structures were observed in the fengycin-treated sample. Arrowheads indicate abnormal morphology of PH-1 hyphae. Bar: 5 µm.

2.3. Fengycin Reduced F. graminearum Pathogenicity and Mycotoxins Biosynthesis

The results of fengycin application showed that fengycin could markedly reduce *F. graminearum* pathogenicity on wheat kernels (Figure 4). As DON is not only an important mycotoxin but also an essential virulence factor produced by *F. graminearum*, we further characterized the effect of fengycin on DON and other mycotoxins, i.e., 3-ADON,15-ADON, and ZEN biosynthesis in *F. graminearum*. The sterilized wheat kernels were incubated with *F. graminearum* and then treated with or without 90 µg/mL purified fengycin, and HPLC–MS analysis was performed to analyze mycotoxins production; we noted that fengycin could noticeably reduce DON, 3-ADON, 15-ADON, and ZEN biosynthesis. (Figure 5). Collectively, these results indicate that fengycin could reduce *F. graminearum* pathogenicity and mycotoxin biosynthesis.

Figure 4. Effects of fengycin on pathogenicity of *F. graminearum*. Wheat heads were drop inoculated with conidial suspensions of *F. graminearum* PH-1 and then were treated with 90 µg/mL fengycin. Conidial suspension with 6.67% (*v/v*) methanol served as the control.

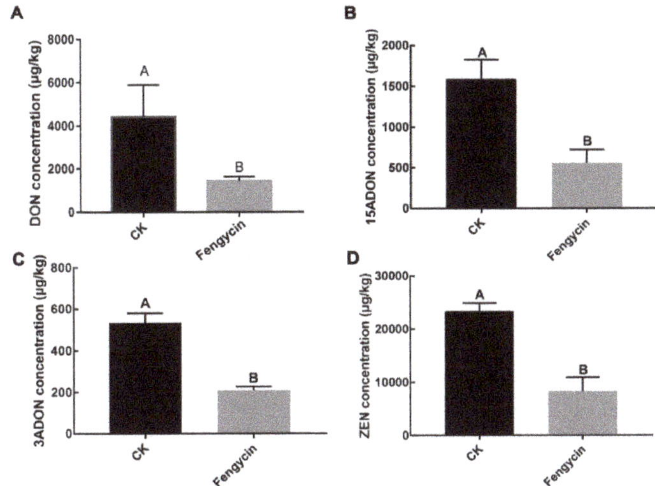

Figure 5. Effects of fengycin (90 µg/mL) on mycotoxins biosynthesis. Amount of DON (**A**), 15-ADON (**B**), 3-ADON (**C**), and ZEN (**D**) in infected wheat kernels 21 days after inoculation, with or without fengycin. Statistical analysis was carried out using Statistix 8.0 and subjected to one-way ANOVA with significant difference detected by Duncan's multiple range test. Line bars denote standard errors of three replicate experiments and different letters describe significant differences at $p < 0.01$ within the same data group.

3. Discussion

Apart from the chemical and conventional methods used for controlling diseases in plants, biological control is one of the safest and most effective alternative methods. Various microorganisms have been reported to be effectively used as a part of this method [15]. For example, *Bacillus* spp. act as effective antifungal and biocontrol organisms [16–18]. They produce many antifungal compounds, such as β-1,3-1,4-glucanase, chitinase, and lipopeptides [19–21]; such compounds can reportedly reduce or demine the activity of various phytopathogenic organisms.

Based on their structures, cyclic lipopeptides can generally be classified into four major families or groups: surfactin, iturin, fengycin, and locillomycin [22,23]. Fengycin, which has been isolated from members of the *Bacillus* genus, demonstrates strong antifungal activity and inhibits the growth of several plant pathogens, particularly of many filamentous fungi [4,24]. Likewise, fengycin antagonistically affected the growth of *F. graminearum* (FG-PH1) in vitro. *B. subtilis* has achieved a lot of attention as a biocontrol agent for manipulating many soil-borne diseases [25]. In this study, fengycin extracted from *B. amyloliquefaciens* FZB42 played a vital role in inhibiting growth of pathogenic fungus *F. graminearum*. Our studies have demonstrated that fengycin has an adverse effect on the structure of fungal hyphae and related activity of *F. graminearum* markedly decreases upon fengycin treatment [19,26–28]. The possible mechanism for the antifungal activity of fengycin is that it interacts with sterol and phospholipid molecules in the fungal cell membrane, altering its structure and permeability [29]. Our results, and even those of some previous studies, suggest that fengycin severely damages the plasma membranes and cell walls of *F. graminearum* hyphae and conidia, consequently causing cell death Fengycin influenced the cell membrane and cellular organs and also inhibited DNA synthesis [30,31]. It can also reportedly cause cell lysis of *M. fructicola* [31,32].

Current studies revealed that fengycin influences the pathogenicity of *F. graminearum* in plants. Fengycin caused a significant reduction in *F. graminearum* virulence, relative to the controls; the lengths of lesions on wheat spikes caused by *F. graminearum* markedly reduced upon treatment with fengycin. Similar results were also obtained in our previous studies when corn silk was treated with bacillomycin D [13].

Sporulation and germination play an important role in the asexual life cycle of *F. graminearum* and in spreading of FHB. In this study, fengycin strongly inhibited both the formation and germination of spores. Fengycin damaged the conidia and inhibited conidial germination; when spores of *F. graminearum* were used as inoculum to infect wheat heads, the severity of disease symptoms was reduced. In addition, considering the adverse effects of fengycin on fungal hyphae, it can inhibit the infection of corn silk by *F. graminearum* hyphae, as reported by a previous study [13]. Hence, the use of microorganisms producing antifungal compounds that inhibit conidial germination and hyphal growth should be an effective measure for biocontrol of fungal diseases.

Mitogen-activated protein kinase (MAPK) signaling pathways have been well characterized in *F. graminearum* [33,34]. Phosphorylation of the MAPKs FgHOG1 and FgMGV1 positively regulates environmental stress responses and DON biosynthesis [35,36]. Here, we observed that 90 µg/mL fengycin was enough to decrease mycotoxin biosynthesis in *F. graminearum*. In contradiction to former results, fengycin reduced the biosynthesis of secondary metabolites in *Fusarium graminearum*, while previous studies showed that the lipopeptides, such as Bacillomycin D and surfactin, induce significant fumonisin production in *F. verticillioides* [37]. Moreover, iturins reportedly induce HOG1 activation in *V. dahliae* [32]. Our results were similar to Kim et al., showing that the *B. amyloliquefaciens* JCK-12 has the ability to decrease both fungal growth and mycotoxin production. CLPs successfully inhibited DON production by affecting DON biosynthetic gene expression [38]. Our results also showed that the fengycin reduces the *F. graminearum* pathogenicity, as well as the most important virulence factor and mycotoxin biosynthesis in *Fusarium graminearum*. In summary, we report that fengycin produced by *Bacillus amyloliquefaciens* FZB42 has potential as a bio-control agent against wheat pathogen *Fusarium graminearum*. For future studies, other important derivatives of DON, such as deepoxy-deoxynivalenol, 3-epi-deoxynivalenol, or deoxynivalenol-3-glucoside, could also be studied under the influence of fengycin.

4. Materials and Methods

4.1. Bacterial and Fungal Strains Growth Conditions

In this study two previously constructed mutants, AK2 (Cannot synthesize fengycin) and AK1S (Can synthesize fengycin), from FZB42, were used [13] for activity against *F. graminearum* PH-1.

Solidified Luria–Bertani medium was used to culture the mutants [39] and were activated in Landy medium [40]. Antibiotics ampicillin, chloramphenicol, and erythromycin were added at the following final concentrations: 100 g/mL, 5 g/mL, and 10 g/mL, respectively. For conidia formation fresh mycelia of PH-1 (50 mg) was taken from the margin of a 72 h grown colony and inoculated in 20 mL mung bean liquid medium [13], followed by incubation at 25 °C for 4 days in a shaker (180 rpm). A hemocytometer was used to count the number of conidia in each flask.

4.2. Anti-fungal Activity Assay

Antifungal activities of AK1S, AK2, and their secondary metabolite extracts were analyzed. Briefly, potato dextrose agar (PDA) was used to measure antifungal activity. A fungal block of 6-mm-diameter was patched at the middle of a PDA plate; 5 μl bacterial suspension (optical density at 600 nm of 2) or their secondary metabolite extracts were inoculated 3 cm away from the fungal block, and the plates were kept at 25 °C for 48 h; later, the diameters of the inhibition zones were measured.

4.3. Purification of Fengycin from AK1S and MALDI-TOF-MS Analysis

Fengycin was purified from AK1S as it could only produce fengycin; there was no possibility of contamination by either surfactin or bacillomycin D. A single colony of AK1S was inoculated into 20 mL LB medium and incubated for 18 h at 37 °C. Six milliliters of this culture was then inoculated in a 500-mL flask containing 200 mL of Landy medium, followed by incubation for 2 days at 30 °C. The sample was then centrifuged at 12,000× g for 20 min at 4 °C with Beckman Coulter Avanti J-26S XP centrifuge with JA-10 rotor (Beckman Coulter Brea, CA, USA) and the supernatant was collected. The pH of this cell-free extract was adjusted to 2, followed by centrifugation at 12,000× g for 20 min at 4 °C. This finally resulted in the precipitation of lipopeptides in the supernatant. Methanol was used to re-dissolve the precipitate, and the pH was adjusted to 7.0 using 1.0 M NaOH [41]. The supernatant was then passed through a silica gel column using different ratios of methanol and methylene chloride, termed MIX1 to MIX3 (methanol/methylene chloride ratios in MIX1, MIX2, and MIX3 were 1:2, 3:1, and 5:1, respectively). Preparative HPLC was performed for fengycin purification, in which we used a microbore 1100 HPLC system (Agilent Technologies, Santa Clara, CA, USA) with a VP 250/21 Nucleodur C18 HTec 5 μm column (Macherey-Nagel, Amtsgericht Düren, Germany); the eluates from MIX2 and MIX3 were collected and further used as mobile phase A, comprised of Acetonitrile with 0.1% (*v/v*) trifluoroacetic acid, and Milli-Q water with 0.1% (*v/v*) trifluoroacetic acid was used as mobile phase B. A solvent containing 45% mobile phase A and 55% mobile phase B was used at a flow rate of 8 mL/min for purification. For detection, UV absorption at 207 nm was recorded. The purity of fengycin collected from different peaks was detected using a 1200 HPLC system (Agilent Technologies, Santa Clara, CA, USA) with an Agilent Eclipse XDB-C18 5 μm column. Fengycin was confirmed by the appearance of one large peak at running times between 10 to 40 min. Purity of fengycin was calculated on the bases of the peak area (96.6%), which was used in our further studies. For antifungal activity of elution components from different peaks, they were tested and analyzed by MALDI-TOF-MS using a Bruker Daltonik Reflex MALDI-TOF instrument with a 337-nm nitrogen laser for desorption and ionization [42]. The α-Cyano-4-hydroxycinnamic acid served as the matrix.

4.4. Scanning Electron Microscopic Observation of Hyphal Morphologies

SEM was used to observe the morphological changes in *F. graminearum* hyphae caused by fengycin (90 μg/mL). To observe the fungal hyphae, they were treated with fengycin and 2.5% glutaraldehyde was used for prefixing. Subsequently, 100 mM phosphate buffer was used to rinse the fixed cells three times for 10 min; the samples were then post-fixed in 1% osmium tetroxide for 3 h and dehydration was done by using an ethanol gradient. Later, the samples were coated with gold particles and electron micrographs were obtained by using a Hitachi Science System Hitachi S-3000N scanning electron microscope at voltage 20 kV (H-7650, Hitachi, 251 Tokyo, Japan).

4.5. Plant Infection and Mycotoxin Production Assay

Wheat spikes were used to assess whether fengycin could adversely affect the pathogenicity of *F. graminearum*. When wheat reached the anthesis stage, the sixth spikelet from the base of the spike was pointed and inoculated with 10 µL of conidial suspension (10^5 conidia/mL) containing 90 µg/mL fengycin. Only 10µl conidial suspension with 6.67% (*v/v*) methanol was used as the control. Three wheat spikes were inoculated for each treatment. The wheat plants were kept at 22 ± 2 °C and under 100% humidity for 2 days and then continued in a glass house. Diseased wheat kernels were examined and counted after 14 days. In order to ascertain the results, the experiment was repeated three times.

For the mycotoxin production assay, 50 g of healthy wheat spikes (wet weight) were surface sterilized by washing with 2% sodium hypochlorite and then inoculated with 1 mL of conidial suspension (10^6 conidia/mL) containing 90 µg/mL fengycin and 6.67% (*v/v*) methanol, and incubated at 25 °C for 20 days. One milliliter of conidial suspension (10^6 conidia/mL) with 6.67% (*v/v*) methanol served as the control. The experiment was repeated three times with three replicates for each. Mycotoxin extraction and quantification was performed by MycoSep 225 Trich Push Columns (Romer Laboratories Diagnostic (Getzersdorf, Austria) according to the manufacturer's guidelines, avoiding the grinding of the starting material. The sample extract residue was dissolved in 400mL methanol/water (30:70). DON was quantified by HPLC according to the protocol followed by [43] with minor modifications. Separation was performed at room temperature by using a C18 reverse α-phase column (120 Å, 5 µm particle size, 4.66 × 150 mm, Acclaim) with an isocratic mobile phase of methanol/water (30:70) at a flow rate of 0.7 mL/min. Eluates were detected using a UV detector set at 220 nm. For quantification of DON, 3-DON, 15-Don, and ZEN, known amounts of pure standard bought from sigma were used as internal standards. Different concentrations of pure compound were analyzed by HPLC, and then the equation depicting the relationship between the concentration and HPLC peak area was obtained, which correlated the peak area to mycotoxin concentration [43].

Author Contributions: A.H. performed all the studies and conducted experiments. Y.X. and M.Z. (Muhammad Zubair) compiled the data and results and A.F., T.M. and J.Y. helped in helped in writing the manuscript. H.W. and J.L. helped with analyzing the data. F.Z., P.L., and C.L. carried out MALDI-TOF MS analyses. X.Z., M.Z. (Mengxuan Zhang), and D.J. performed HPLC analyses; Q.G. and X.G. supervised the whole of the work and the revision of the manuscript.

Funding: This work was supported by the Natural Science Foundation of Jiangsu Province, China [grant number BK20160719] and the National Natural Science Foundation of China [grant number 31601589].

Acknowledgments: We would like to extend our gratitude towards native English-speaking scientists of Elixigen Company (Huntington Beach, CA, USA) for editing our manuscript.

Conflicts of Interest: The authors declare no conflict of interest.

References

1. Goswami, R.S.; Kistler, H.C. Heading for disaster: *Fusarium graminearum* on cereal crops. *Mol. Plant Pathol.* **2004**, *5*, 515–525. [CrossRef] [PubMed]

2. Yu, C.; Wang, W.-X.; Zhang, A.-F.; Gu, C.-Y.; Zhou, M.-G.; Gao, T.-C. Activity of the fungicide JS399-19 against *Fusarium* head blight of wheat and the risk of resistance. *Agr. Sci. China* **2011**, *10*, 1906–1913.

3. Bily, A.C.; Reid, L.M.; Savard, M.E.; Reddy, R.; Blackwell, B.A.; Campbell, C.M.; Krantis, A.; Durst, T.; Philogene, B.J.; Arnason, J.T.; et al. Analysis of *Fusarium graminearum* mycotoxins in different biological matrices by LC/MS. *Mycopathologia* **2004**, *157*, 117–126. [CrossRef] [PubMed]

4. Guo, Q.; Dong, W.; Li, S.; Lu, X.; Wang, P.; Zhang, X.; Wang, Y.; Ma, P. Fengycin produced by *Bacillus subtilis* NCD-2 plays a major role in biocontrol of cotton seedling damping-off disease. *Microbiol. Res.* **2014**, *169*, 533–540. [CrossRef]

5. Voss, K.A. A new perspective on deoxynivalenol and growth suppression. *Toxicol. Sci.* **2010**, *113*, 281–283. [CrossRef]

6. Parry, D.; Jenkinson, P.; McLeod, L. *Fusarium* ear blight (scab) in small grain cereals—A review. *Plant Pathol.* **1995**, *44*, 207–238.

7. Wang, J.; Liu, J.; Chen, H.; Yao, J. Characterization of *Fusarium graminearum* inhibitory lipopeptide from *Bacillus subtilis* IB. *Appl. Microbiol. Biotechnol.* **2007**, *76*, 889–894. [CrossRef]

8. Chen, X.H.; Koumoutsi, A.; Scholz, R.; Eisenreich, A.; Schneider, K.; Heinemeyer, I.; Morgenstern, B.; Voss, B.; Hess, W.R.; Reva, O. Comparative analysis of the complete genome sequence of the plant growth–promoting bacterium *Bacillus amyloliquefaciens* FZB42. *Nat. Biotechnol.* **2007**, *25*, 1007. [CrossRef]

9. Lugtenberg, B.; Kamilova, F. Plant-growth-promoting rhizobacteria. *Annu. Rev. Microbiol.* **2009**, *63*, 541–556. [CrossRef]

10. Koumoutsi, A.; Chen, X.-H.; Vater, J.; Borriss, R. DegU and YczE positively regulate the synthesis of bacillomycin D by *Bacillus amyloliquefaciens* strain FZB42. *Appl. Environ. Microbiol.* **2007**, *73*, 6953–6964. [CrossRef]

11. Bulgarelli, D.; Schlaeppi, K.; Spaepen, S.; van Themaat, E.V.L.; Schulze-Lefert, P. Structure and functions of the bacterial microbiota of plants. *Annu. Rev. Plant Biol.* **2013**, *64*, 807–838. [CrossRef]

12. Lugtenberg, B.J.; Dekkers, L.; Bloemberg, G.V. Molecular determinants of rhizosphere colonization by *Pseudomonas. Annu. Rev. Plant Biol.* **2001**, *39*, 461–490. [CrossRef]

13. Gu, Q.; Yang, Y.; Yuan, Q.; Shi, G.; Wu, L.; Lou, Z.; Huo, R.; Wu, H.; Borriss, R.; Gao, X. Bacillomycin D Produced by *Bacillus amyloliquefaciens* Is Involved in the Antagonistic Interaction with the Plant-Pathogenic Fungus *Fusarium graminearum. Appl. Environ. Microbiol.* **2017**, *83*. [CrossRef] [PubMed]

14. Epstein, A.K.; Pokroy, B.; Seminara, A.; Aizenberg, J. Bacterial biofilm shows persistent resistance to liquid wetting and gas penetration. *Proc. Nat. Acad. Sci.* **2011**, *108*, 995–1000. [CrossRef]

15. Niu, B.; Vater, J.; Rueckert, C.; Blom, J.; Lehmann, M.; Ru, J.-J.; Chen, X.-H.; Wang, Q.; Borriss, R. Polymyxin P is the active principle in suppressing phytopathogenic *Erwinia* spp. by the biocontrol rhizobacterium *Paenibacillus polymyxa* M-1. *BMC Microbiol.* **2013**, *13*, 137. [CrossRef] [PubMed]

16. Chen, X.; Zhang, Y.; Fu, X.; Li, Y.; Wang, Q. Isolation and characterization of *Bacillus amyloliquefaciens* PG12 for the biological control of apple ring rot. *Postharvest Biol. Tec.* **2016**, *115*, 113–121. [CrossRef]

17. JI, Z.-l.; LING, Z.; ZHANG, Q.-x.; XU, J.-y.; CHEN, X.-j.; TONG, Y.-h. Study on the inhibition of *Bacillus licheniformis* on *Botryosphaeria berengeriana* f. sp. piricola and *Glomerella cingulata* and biocontrol efficacy on postharvest apple diseases. *J. Fruit Sci.* **2008**, *2*, 019.

18. Li, Y.; Han, L.-R.; Zhang, Y.; Fu, X.; Chen, X.; Zhang, L.; Mei, R.; Wang, Q. Biological control of apple ring rot on fruit by *Bacillus amyloliquefaciens* 9001. *Plant Pathol. J.* **2013**, *29*, 168. [CrossRef] [PubMed]

19. Alvarez, F.; Castro, M.; Principe, A.; Borioli, G.; Fischer, S.; Mori, G.; Jofre, E. The plant-associated *Bacillus amyloliquefaciens* strains MEP2 18 and ARP2 3 capable of producing the cyclic lipopeptides iturin or surfactin and fengycin are effective in biocontrol of *sclerotinia* stem rot disease. *J. Appl. Microbiol.* **2012**, *112*, 159–174. [CrossRef]

20. Xu, Y.-B.; Chen, M.; Zhang, Y.; Wang, M.; Wang, Y.; Huang, Q.-b.; Wang, X.; Wang, G. The phosphotransferase system gene ptsI in the endophytic bacterium *Bacillus cereus* is required for biofilm formation, colonization, and biocontrol against wheat sharp eyespot. *FEMS Microbiol. Lett.* **2014**, *354*, 142–152. [CrossRef]

21. Zhang, Q.; Yong, D.; Zhang, Y.; Shi, X.; Li, B.; Li, G.; Liang, W.; Wang, C. Streptomyces rochei A-1 induces resistance and defense-related responses against *Botryosphaeria dothidea* in apple fruit during storage. *Postharvest Biol. Tec.* **2016**, *115*, 30–37. [CrossRef]

22. Luo, C.; Liu, X.; Zhou, H.; Wang, X.; Chen, Z. Nonribosomal peptide synthase gene clusters for lipopeptide biosynthesis in *Bacillus subtilis* 916 and their phenotypic functions. *Appl. Environ. Microbiol.* **2015**, *81*, 422–431. [CrossRef]

23. Ongena, M.; Jacques, P. *Bacillus* lipopeptides: Versatile weapons for plant disease biocontrol. *Trends Microbiol.* **2008**, *16*, 115–125. [CrossRef]

24. Afsharmanesh, H.; Ahmadzadeh, M.; Javan-Nikkhah, M.; Behboudi, K. Improvement in biocontrol activity of *Bacillus subtilis* UTB1 against *Aspergillus flavus* using gamma-irradiation. *Crop Prot.* **2014**, *60*, 83–92. [CrossRef]

25. Leclère, V.; Béchet, M.; Adam, A.; Guez, J.-S.; Wathelet, B.; Ongena, M.; Thonart, P.; Gancel, F.; Chollet-Imbert, M.; Jacques, P. Mycosubtilin Overproduction by *Bacillus subtilis* BBG100 Enhances the Organism's Antagonistic and Biocontrol Activities. *Appl. Environ. Microbiol.* **2005**, *71*, 4577–4584. [CrossRef] [PubMed]

26. Falardeau, J.; Wise, C.; Novitsky, L.; Avis, T.J. Ecological and mechanistic insights into the direct and indirect antimicrobial properties of *Bacillus subtilis* lipopeptides on plant pathogens. *J. Chem. Ecol.* **2013**, *39*, 869–878. [CrossRef]

27. Roy, A.; Mahata, D.; Paul, D.; Korpole, S.; Franco, O.L.; Mandal, S.M. Purification, biochemical characterization and self-assembled structure of a fengycin-like antifungal peptide from *Bacillus thuringiensis* strain SM1. *Front. Microbiol.* **2013**, *4*, 332. [CrossRef]
28. Tang, Q.; Bie, X.; Lu, Z.; Lv, F.; Tao, Y.; Qu, X. Effects of fengycin from *Bacillus subtilis* fmbJ on apoptosis and necrosis in *Rhizopus stolonifer*. *J. Microbiol.* **2014**, *52*, 675–680. [CrossRef]
29. Deleu, M.; Paquot, M.; Nylander, T. Fengycin interaction with lipid monolayers at the air–aqueous interface—Implications for the effect of fengycin on biological membranes. *J. Colloid Interf. Sci.* **2005**, *283*, 358–365. [CrossRef]
30. Tao, Y.; Bie, X.M.; Lv, F.X.; Zhao, H.Z.; Lu, Z.X. Antifungal activity and mechanism of fengycin in the presence and absence of commercial surfactin against *Rhizopus stolonifer*. *J. Microbiol.* **2011**, *49*, 146–150. [CrossRef] [PubMed]
31. Liu, J.; Zhou, T.; He, D.; Li, X.-z.; Wu, H.; Liu, W.; Gao, X. Functions of lipopeptides bacillomycin D and fengycin in antagonism of *Bacillus amyloliquefaciens* C06 towards *Monilinia fructicola*. *J. Mol. Microbiol. Biotechnol.* **2011**, *20*, 43–52. [CrossRef] [PubMed]
32. Han, Q.; Wu, F.; Wang, X.; Qi, H.; Shi, L.; Ren, A.; Liu, Q.; Zhao, M.; Tang, C. The bacterial lipopeptide iturins induce *Verticillium dahliae* cell death by affecting fungal signalling pathways and mediate plant defence responses involved in pathogen-associated molecular pattern-triggered immunity. *Environ. Microbiol.* **2015**, *17*, 1166–1188. [CrossRef] [PubMed]
33. Wang, C.; Zhang, S.; Hou, R.; Zhao, Z.; Zheng, Q.; Xu, Q.; Zheng, D.; Wang, G.; Liu, H.; Gao, X. Functional analysis of the kinome of the wheat scab fungus *Fusarium graminearum*. *PLoS Pathog.* **2011**, *7*, e1002460. [CrossRef]
34. Gu, Q.; Zhang, C.; Yu, F.; Yin, Y.; Shim, W.B.; Ma, Z. Protein kinase FgSch 9 serves as a mediator of the target of rapamycin and high osmolarity glycerol pathways and regulates multiple stress responses and secondary metabolism in *Fusarium graminearum*. *Environ. Microbiol.* **2015**, *17*, 2661–2676. [CrossRef]
35. Zheng, D.; Zhang, S.; Zhou, X.; Wang, C.; Xiang, P.; Zheng, Q.; Xu, J.-R. The FgHOG1 pathway regulates hyphal growth, stress responses, and plant infection in *Fusarium graminearum*. *PloS ONE* **2012**, *7*, e49495. [CrossRef]
36. Hou, Z.; Xue, C.; Peng, Y.; Katan, T.; Kistler, H.C.; Xu, J.-R. A mitogen-activated protein kinase gene (MGV1) in *Fusarium graminearum* is required for female fertility, heterokaryon formation, and plant infection. *Mol. Plant Microbe Interact.* **2002**, *15*, 1119–1127. [CrossRef] [PubMed]
37. Blacutt, A.; Mitchell, T.; Bacon, C.; Gold, S. *Bacillus mojavensis* RRC101 lipopeptides provoke physiological and metabolic changes during antagonism against *Fusarium verticillioides*. *Mol. Plant Microbe Interact.* **2016**, *29*, 713–723. [CrossRef]
38. Kim, K.; Lee, Y.; Ha, A.; Kim, J.-I.; Park, A.R.; Yu, N.H.; Son, H.; Choi, G.J.; Park, H.W.; Lee, C.W.; et al. Chemosensitization of *Fusarium graminearum* to Chemical Fungicides Using Cyclic Lipopeptides Produced by *Bacillus amyloliquefaciens* Strain JCK-12. *Front. Plant Sci.* **2017**, *8*. [CrossRef] [PubMed]
39. Moyne, A.L.; Shelby, R.; Cleveland, T.E.; Tuzun, S. Bacillomycin D: an iturin with antifungal activity against Aspergillus flavus. *Journal of applied microbiology* **2001**, *90*, 622–629. [CrossRef]
40. Koumoutsi, A.; Vater, J.; Junge, H.; Krebs, B.; Borriss, R. Sequence for the Bacillomycin D synthesis in Bacillus amyloliquefaciens FZB42. WO2004111240A2, 23 December 2004.
41. 41 Lin, S.C.; Minton, M.A.; Sharma, M.M.; Georgiou, G. Structural and immunological characterization of a biosurfactant produced by *Bacillus licheniformis* JF-2. *Appl. Environ. Microbiol.* **1994**, *60*, 31–38.
42. Vater, J.; Gao, X.; Hitzeroth, G.; Wilde, C.; Franke, P. "Whole cell"-matrix-assisted laser desorption ionization-time of flight-mass spectrometry, an emerging technique for efficient screening of biocombinatorial libraries of natural compounds-present state of research. *Com. Chem. High Throughput Screen.* **2003**, *6*, 557–567. [CrossRef]
43. Omurtag, G.Z.; Beyoğlu, D. Occurrence of deoxynivalenol (vomitoxin) in beer in Turkey detected by HPLC. *Food Control* **2007**, *18*, 163–166. [CrossRef]

Article

Effects of Essential Oil Citral on the Growth, Mycotoxin Biosynthesis and Transcriptomic Profile of *Alternaria alternata*

Liuqing Wang [1,2], Nan Jiang [1,2], Duo Wang [1,2] and Meng Wang [1,2,*]

[1] Beijing Research Center for Agricultural Standards and Testing, No. 9 Middle Road of Shuguanghuayuan, Haidian District, Beijing 100097, China; wanglq@brcast.org.cn (L.W.); jiangn@brcast.org.cn (N.J.); wangduo@brcast.org.cn (D.W.)

[2] Laboratory of Quality & Safety Risk Assessment for Agro-products (Beijing), Ministry of Agriculture and Rural Affairs, No. 9 Middle Road of Shuguanghuayuan, Haidian District, Beijing 100097, China

* Correspondence: wangm@brcast.org.cn; Tel.: +86-10-5150-3178

Received: 27 August 2019; Accepted: 16 September 2019; Published: 20 September 2019

Abstract: *Alternaria alternata* is a critical phytopathogen that causes foodborne spoilage and produces a polyketide mycotoxin, alternariol (AOH), and its derivative, alternariol monomethyl ether (AME). In this study, the inhibitory effects of the essential oil citral on the fungal growth and mycotoxin production of *A. alternata* were evaluated. Our findings indicated that 0.25 µL/mL (222.5 µg/mL) of citral completely suppressed mycelial growth as the minimum inhibitory concentration (MIC). Moreover, the 1/2MIC of citral could inhibit more than 97% of the mycotoxin amount. Transcriptomic profiling was performed by comparative RNA-Seq analysis of *A. alternata* with or without citral treatment. Out of a total of 1334 differentially expressed genes (DEGs), 621 up-regulated and 713 down-regulated genes were identified under citral stress conditions. Numerous DEGs for cell survival, involved in ribosome and nucleolus biogenesis, RNA processing and metabolic processes, and protein processing, were highly expressed in response to citral. However, a number of DEGs responsible for the metabolism of several carbohydrates and amino acids, sulfate and glutathione metabolism, the metabolism of xenobiotics and transporter activity were significantly more likely to be down-regulated. Citral induced the disturbance of cell integrity through the disorder of gene expression, which was further confirmed by the fact that exposure to citral caused irreversibly deleterious disruption of fungal spores and the inhibition of ergosterol biosynthesis. Citral perturbed the balance of oxidative stress, which was likewise verified by a reduction of total antioxidative capacity. In addition, citral was able to modulate the down-regulation of mycotoxin biosynthetic genes, including *pksI* and *omtI*. The results provide new insights for exploring inhibitory mechanisms and indicate citral as a potential antifungal and antimytoxigenic alternative for cereal storage.

Keywords: *Alternaria alternata*; mycotoxin; alternariol; essential oil; cell integrity; oxidative stress

Key Contribution: Citral significantly impaired mycelial growth and mycotoxin biosynthesis in a positive dose-response relationship. Citral stress caused oxidative imbalance, cell integrity disruption, transporter repression as well as the down-regulation of AOH and AME biosynthetic genes, such as *pksI* and *omtI*, as evidenced by transcriptomic profiling.

---

1. Introduction

Alternaria alternata is a widespread phytopathogen, causing serious foodborne spoilage and producing a variety of mycotoxins that are detrimental to human and animal health through food chains. *Alternaria* mycotoxins have gradually been paid more attention in relation to public health, as suggested by the European Food Safety Authority (EFSA) [1]. The major *Alternaria* mycotoxins are alternariol (AOH), alternariol monomethyl ether (AME), altenuene, tenuazonic acid, and altertoxins [2,3]. AOH and AME are two of the most frequent contaminants in food and feedstuffs derived from cereals, fruits and vegetables [2,4]. AOH and AME exert genotoxicity and mutagenicity as topoisomerase poison inducing DNA strand breaks [5], and further they are a possible factor in human oesophageal cancer in China [6,7]. Besides, they possess cytotoxicity as well as reproductive and developmental toxicity [4]. In addition, *A. alternata* is regarded as a critically common cause of allergic rhinitis and atopic asthma owing to the production of *Alternaria* allergens, such as the major allergen, Alt a 1 [8,9].

To date, the AOH and AME biosynthetic pathway has been unraveled, and a polyketide gene cluster is responsible for their production in the genome of *A. alternata* [10]. Among the clustered genes, *pksI* encoding for polyketide synthase was sufficient for AOH biosynthesis, which was verified by gene disruption in *A. alternata* and heterologous expression in *Aspergillus oryzae*. *OmtI* encoding for O-methyl transferase was responsible for AME formation from the methyl ether of AOH. Except for *omtI*, another three clustered genes, encoding a mono-oxygenase (*moxI*), a short-chain dehydrogenase (*sdrI*) and an estradiol dioxygenase (*doxI*), were also involved in AOH modification. In this cluster, a fungal specific transcriptional factor encoded by *aohR* positively regulated the expression of other clustered genes. In addition, *pksI* homologous gene, *SnPKS19*, was illustrated to be involved in AOH biosynthesis in the wheat pathogen *Parastagonospora nodorum*, and likewise this was further proved by heterologous expression in *A. nidulans* [11]. Moreover, AOH and AME demonstrated potential phytotoxic activity as pathogenicity factors in their study. AOH facilitated the infection and colonization of *A. alternata* on tomato, citrus and apple by the addition of AOH to *pksI*-deletion mutant, but the chemical alone exhibited no phytotoxin in wheat leaves and seed germination [10,11]. Correspondingly, AME enabled the inhibition of photosynthetic electron transport in extracted spinach chloroplasts [12].

Essential oils are easily volatile aromatic compounds extracted from plant material, including terpenes, aldehydes, esters, etc. They are environmentally friendly and are employed as ingredients in drugs, cosmetics, etc. The components of essential oils, such as citral, cinnamaldehyde, eugenol and thymol, exhibit strong antifungal properties as well as the inhibition of mycotoxin production [13–15]. Among them, citral has broad-spectrum inhibitory effects against various plant pathogens, including *A. solani* [16], *Penicillium italicum* [17–19], *P. expansum* [20], *A. flavus*, *Fusarium moniliforme*, etc. [21]. Furthermore, citral displayed cytotoxicity against *P. italicum* by affecting mitochondrial dysfunction, damaging membranes and inhibiting ergosterol biosynthesis in previous studies [17–19]. Moreover, the combination of citral and cinnamaldehyde highly suppressed the growth and patulin production of *P. expansum* by oxidative damage and down-regulation of the patulin biosynthetic pathway [20]. In addition to the antifungal activity of *A. flavus*, citral exerted antiaflatoxigenic activity by modulating aflatoxin biosynthetic gene expression [22,23].

Citral revealed fungitoxic activity on *A. alternata* by a paper disc agar diffusion assay, as suggested by Kishore et al. [21]. However, until now, citral's mode of action in repressing growth and mycotoxin production in *A. alternata* has not yet been elucidated. In this study, the inhibitory effects of citral on fungal growth and mycotoxin biosynthesis were evaluated in *A. alternata*. Furthermore, great efforts were made to uncover the potential mechanisms through a comprehensive and systematical view by RNA-Seq among samples with or without citral treatment. Our findings offer promising insights into citral's application for the control of fungal infection and mycotoxin contamination.

2. Results

2.1. Inhibitory Effects on Mycelial Growth, Spore Germination and Mycotoxin Production

The inhibitory effects of citral against *A. alternata* are displayed in Figure 1. The mycelial growth was significantly inhibited in a dose-dependent manner (Figure 1A). The inhibition percentage rose from 21.3% to 47.5% with the increase in citral concentration from 0.0625 to 0.125 µL/mL (from 55.625 to 111.25 µg/mL). The mycelial growth of *A. alternata* was completely suppressed at 0.25 µL/mL

(222.5 μg/mL) as the minimum inhibitory concentration (MIC). Moreover, the minimum fungicidal concentration (MFC) of citral was determined to be 1.0 μL/mL (890 μg/mL) by the observation of no mycelial growth on the citral free culture after citral treatment.

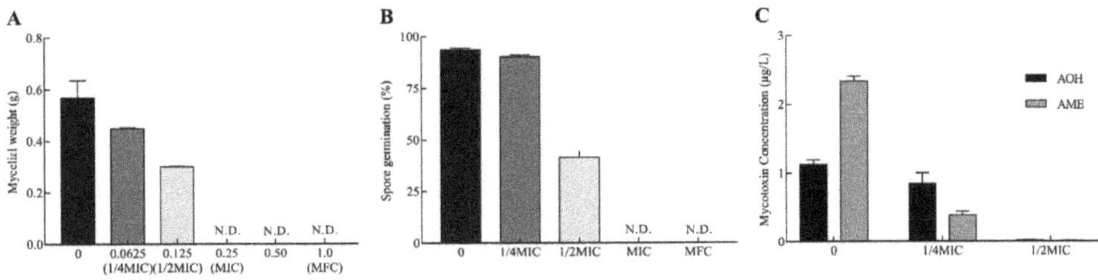

Figure 1. Citral affects mycelial weight, spore germination and mycotoxin production of *A. alternata*. (**A**) The mycelial weight of *A. alternata* was determined after drying under serial concentrations of citral. Minimum inhibitory concentration (MIC): 0.25 μL/mL (222.5 μg/mL); minimum fungicidal concentration (MFC): 1.0 μL/mL (890 μg/mL). (**B**) The germination rate of *A. alternata* spores exposed to citral. (**C**) The determination of mycotoxins, alternariol (AOH) and alternariol monomethyl ether (AME) produced by *A. alternata* in response to citral. The results are illustrated as mean ± SEM ($n = 3$).

Likewise, spore germination was markedly repressed in response to different concentrations of citral (Figure 1B). The presence of spore germination was almost entirely observed in the control by light microscopy, while the absence of spore germination was observed at the MIC and MFC of citral. Therefore, citral revealed outstanding antifungal properties against mycelial growth and spore germination in *A. alternata*.

Citral obviously repressed AOH and AME production in *A. alternata* (Figure 1C). At 1/4MIC of citral, AOH concentration did not sharply reduce in *A. alternata* ($p = 0.19$). However, AOH was hardly biosynthesized by *A. alternata* and declined by 98.6% at 1/2MIC. Citral was markedly effective in resisting against AME production at 1/4MIC and 1/2MIC. The inhibition rates exhibited were 83.3% and 99.6%, respectively, compared with the control. In connection with mycelial growth, the inhibition of mycelial growth was not the only cause of the reduction of the mycotoxin amount.

2.2. Global Analysis of Transcriptomic Profile

Comparative transcriptome analysis of *A. alternata* was performed to systematically discover the potential antifungal and antimycotoxigenic mechanisms between the control and 1/2MIC citral treatment. The total statistics of the RNA-Seq data are summarized in Table S1. More than 48 million clean reads were obtained after removing adaptor sequences, low-quality reads, sequences with 10% higher ambiguous bases, and over-short sequences. The average rate of total mapped reads was about 87.4% after the alignment with the sequence of the reference genome. The results of the correlation between biological replicates were higher than 0.96, which was performed by Pearson's correlation coefficient based on the expression matrix. Out of 13,761 genes in total, Veen analysis showed that there were 10,767 transcribed genes in the control and 10,304 expressed genes in the 1/2MIC citral treatment (Figure 2A,B). It was shown that 627 unique genes were only transcribed in the control, and 164 genes were solely expressed under the citral condition.

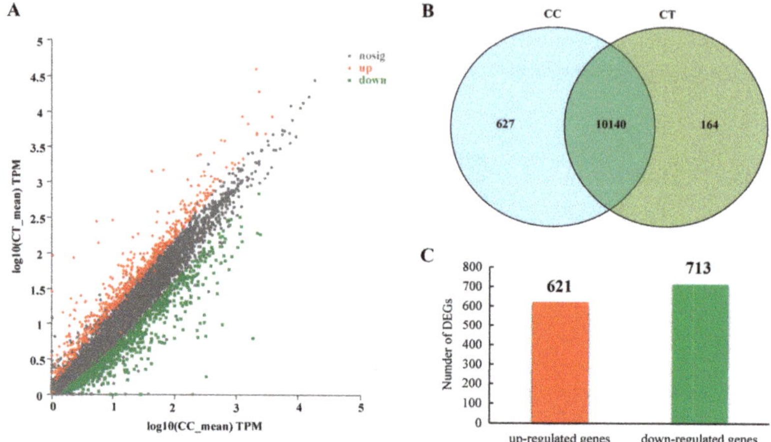

Figure 2. Summary of RNA-Seq analysis. (**A**) Statistical scatter diagram of differential gene expression. Compared to the control of *A. alternata* with no citral (CC), the pattern of gene expression was indicated as up-regulation (up), down-regulation (down) and no significant differential expression (nosig) under the condition of citral treatment (CT). TPM: transcripts per million reads. (**B**) Venn diagram of the transcribed genes between citral-treated and untreated samples. (**C**) Number of differentially expressed genes (DEGs) in *A. alternata* exposed to citral stress. Out of 1334 DEGs, 621 genes were highly up-regulated, and 713 genes were markedly down-regulated.

The analyses of RNA-Seq data involved great efforts to explore the potential differential genes and pathways in *A. alternata* under citral stress. Differentially expressed genes (DEGs) were analyzed and are revealed in Table S2 based on the absolute value of fold change (FC) at ≥2 and false discovery rate (FDR) at < 0.05. In summary, there were 1334 DEGs found within the two groups of citral-treated and untreated fungi, with 621 (47%) up-regulated and 713 (53%) down-regulated DEGs compared to the control (Figure 2A,C).

2.3. Functional Analysis of DEGs

To further characterize the functional differences and relationships of DEGs, they were excavated by Gene Ontology (GO) enrichment and Kyoto Encyclopedia of Genes and Genomes (KEGG) pathway enrichment analyses. Based on the significant results of the GO enrichment analysis, there were obvious differences between GO enrichment analyses separately performed based on the up-regulated or down-regulated DEGs (Figure 3). The up-regulated DEGs in association with the processes of ribosome formation and protein processing were highly represented, including rRNA processing and metabolic processes, ncRNA processing and metabolic processes, RNA processing, preribosome, protein folding and unfolded protein binding, which are essential for fundamental fungal survival during exposure to stress. Nevertheless, the down-regulated DEGs were significantly enriched in oxidoreductase activity, antibiotic metabolic processes, catalytic activity, and transporter activity. The result of gene expression related with cellular RNA behavior was a little different from *P. digitatum* in response to citral stress over a short time [17]. This might reveal that there are some differences in antifungal mechanisms for different fungal species under long- or short-term chemical treatment.

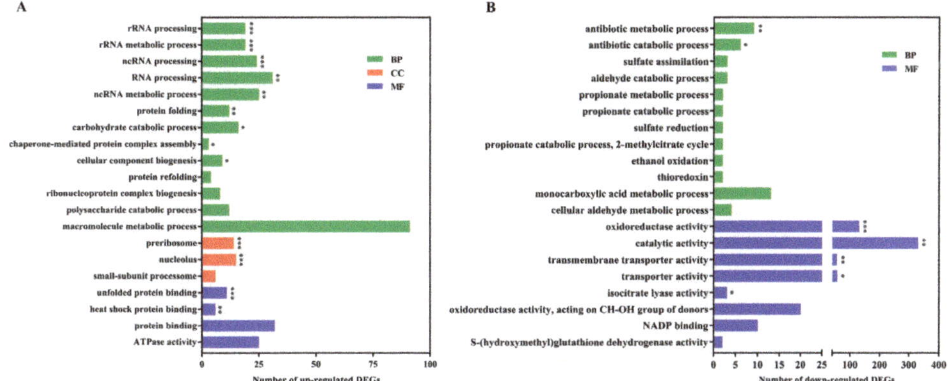

Figure 3. Gene ontology (GO) enrichment analysis of differentially expressed genes (DEGs). Top 20 results of the highest enrichment level were separately obtained from the analysis of up-regulated DEGs (**A**) and down-regulated DEGs (**B**). ***: false discovery rate (FDR) < 0.001; **: FDR < 0.01; *: FDR < 0.05.

To uncover the metabolic pathway of these DEGs, they were mapped into the KEGG pathway database. Ribosome biogenesis in eukaryotes and protein processing in endoplasmic reticulum were the most significant enrichment of pathways in the up-regulated DEGs (Figure 4). However, there were a large number of categories in relation to carbohydrate metabolism, energy metabolism and xenobiotics biodegradation, which were highly represented in the down-regulated DEGs. To a certain extent, a number of fungal processes for primary metabolism, especially carbohydrate and energy metabolism including glyoxylate and dicarboxylate metabolism, pyruvate metabolism glycolysis/gluconeogenesis and methane metabolism, were seriously hampered by citral stress. Correspondingly, among these genes, 11 DEGs were shown to be down-regulated in pyruvate metabolism, and eight DEGs were inhibited in the glycolysis/gluconeogenesis pathway by citral. In addition, the processes of nitrogen metabolism and amino acid metabolism were partially repressed under the citral condition.

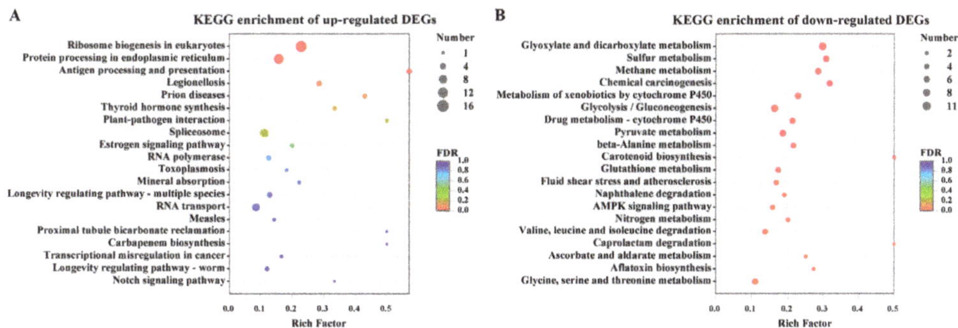

Figure 4. Kyoto Encyclopedia of Genes and Genomes (KEGG) metabolic pathway enrichment analysis of differentially expressed genes (DEGs). Top 20 results of the highest enrichment level were acquired from the individual analysis of up-regulated DEGs (**A**) or down-regulated DEGs (**B**).

2.4. Genes Responsible for Cell Integrity

Cell integrity is essential for fungal survival when exposed to chemical stress. The fungal disruption induces the efflux of cytoplasmic constituents. In this study, the permeability of cytoplasmic constituents was reflected by the leakage of intracellular proteins released from fungal spores of *A. alternata* treated with different concentrations of citral (Figure 5A). Protein release was positively dose-dependent with citral concentration. At the MFC (1.0 μL/mL; 890 μg/mL) of citral, the released protein concentration significantly increased up to 101.7 μg/mL. This showed that cell damage was much more severe in a concentration-dependent way. The permeability was coincident with the disruption of fungal spores observed by microscopic morphology (Figure 5B). The conidia revealed a massive distortion of morphological structure and abnormal cell shrinkage under the MIC of citral. Even worse, there were several completely disrupted spores that occurred in the citral-treated solution.

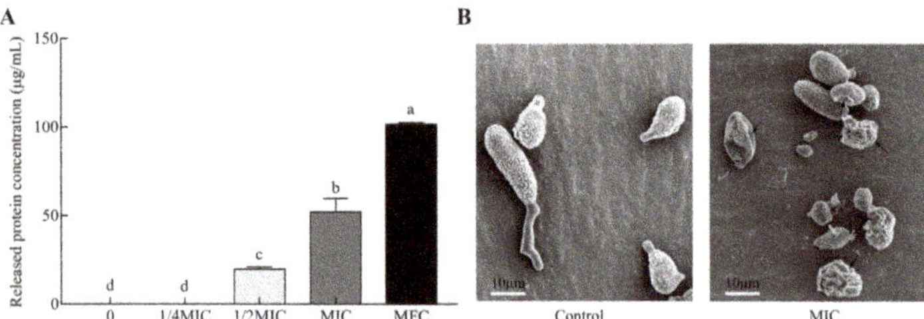

Figure 5. Permeability and morphological alteration of *A. alternata* in response to citral. (**A**) Intracellular protein leakage of *A. alternata* spores treated with serial concentrations of citral ($n = 3$). MIC: minimum inhibitory concentration; MFC: minimum fungicidal concentration. Significant differences ($p < 0.05$) are indicated by different lowercase letters above the bars. (**B**) Images of fungal spores under the conditions of 0 and MIC of citral by scanning electron microscopy.

The fungal cell wall firstly senses the pressure of toxic compounds in the external environment as the outermost defensive line. Two DEGs in relation with cell wall biogenesis and organization were highly up-regulated 3.162- and 6.321-fold in response to citral (Figure 6). Polysaccharides are the principal components for cell wall structure. Of seven DEGs responsible for the polysaccharide catabolic process, except NADP-dependent mannitol dehydrogenase encoding genes, six DEGs were overexpressed between 2.574- and 4.109-fold under the citral condition. In addition, the conserved cell wall integrity pathway is affected by the unfolded protein response in filamentous fungi [24]. In this work, the expression of two genes (*CC77DRAFT_1024747* and *CC77DRAFT_1028832*) involved in the response to unfolded protein was disturbed under chemical stress.

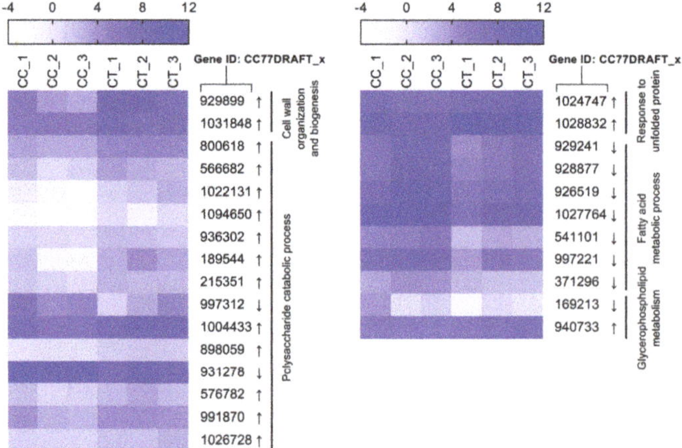

Figure 6. Comparative expression level of DEGs potentially involved in cell integrity. Gene transcriptional values are indicated as $\log_2$ TPM (transcripts per million reads) for each biological replicate of *A. alternata* with (CT) or without (CC) citral treatment. Arrows next to the gene IDs are used to represent up-regulated (↑) and down-regulated (↓) DEGs.

In previous studies, essential oils have been proposed to possess antifungal and antimycotoxigenic potency via the disruption of the plasma membrane as a potential target [25–27]. Ergosterol amount significantly decreased by 36.6% from 3.3 to 2.1 mg/g in response to citral at 1/2MIC compared to the control (Figure S1). Correspondingly, two genes encoding a C-3 sterol dehydrogenase/C-4 decarboxylase-like protein (*ERG26*; *CC77DRAFT_590852*) and a 3-keto-steroid reductase (*ERG27*; *CC77DRAFT_76456*) were differentially expressed in the pathway of ergosterol biosynthesis. However, while the *ERG26* transcriptional level was lower, *ERG27* was more highly transcribed under the stress condition. Virtually, ERG26 catalysate is the substrate of ERG27. A lower intermediate catalyzed by ERG26 was supplied for the following catalysis during the process of ergosterol biosynthesis, in spite of the higher mRNA level of *ERG27*. This might be the reason why ergosterol production drastically declined after citral treatment. Glycerophospholipid plays an important role in the component of the plasma membrane. Glycerophospholipid metabolism was partially interfered with for the differential expressions of *CC77DRAFT_169213* and *CC77DRAFT_940733*. In addition, fatty acid biosynthesis contributed to the fluidity of plasma membranes, and seven DEGs for fatty acid metabolic processes were significantly down-expressed in response to citral, especially two genes of fatty acid synthase activity (*FAS1: CC77DRAFT_928877*; *FAS2: CC77DRAFT_929241*).

2.5. Genes Related to Stress Response

Exposure to citral led to a wide alteration of *A. alternata*'s transcriptomic profile in association with stress response. Essential oils can interfere with the homeostasis of oxidative stress and induce the imbalance of reactive oxygen species (ROS), such as hydrogen peroxide (H_2O_2), superoxide (O_2^-), and hydroxyl radical ($\cdot OH$), during the inhibition of mycotoxin-producing fungi [20,28,29]. Total antioxidant capacity (T-AOC) indeed mirrors the level of ROS balance. In this study, the T-AOC of *A. alternata* was reflected by the ferric reducing ability of the Ferric Reducing Ability of Plasma (FRAP) method and significantly reduced from 0.23 to 0.14 mmol/g with the increase in citral concentration (Figure 7A). This reduction could result from the decreasing content of antioxidant materials scavenging ROS. This might lead to ROS instability and, eventually, the damage of cell structure and secondary metabolism. This was the possible cause of cell debris and the mycotoxin reduction of *A. alternata* suppressed by citral.

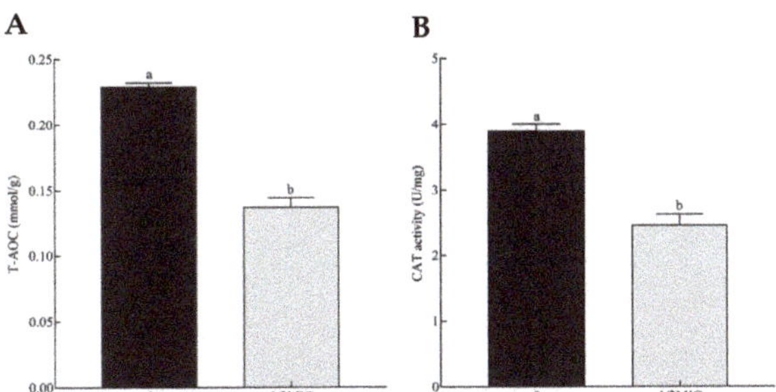

Figure 7. Total antioxidant capacity (**A**) and catalase activity (**B**) of *A. alternata* exposed to 0 and 1/2MIC (minimum inhibitory concentration) of citral. The data are expressed as the mean ± SEM (*n* = 3). Different lowercase letters above the column indicate statistically significant results (*p* < 0.05).

Both enzymatic and non-enzymatic systems are developed for maintaining ROS balance, including cellular detoxifying enzymes and reducing substances. Catalase can catalyze H_2O_2 to detoxify ROS stress. Two isozymes of catalase (CC77DRAFT_364732 and CC77DRAFT_1013212) were significantly down-regulated. Correspondingly, the activity of catalase decreased by 36.9% compared to the control (Figure 7B). In addition, peroxisomes exhibit multifunctional activities, including the decomposition of ROS [30,31]. In this work, seven DEGs responsible for peroxisome biogenesis were less transcribed under the citral condition (Figure 8). This probably resulted in the dysfunction of peroxisome, which was detrimental to the survival of *A. alternata*. Glutathione metabolism responsible for oxidative balance was enriched in down-regulated DEGs from KEGG analysis. Glutathione S-transferase, belonging to the glutathione system, catalyzes the conjugation between glutathione and many xenobiotic compounds for the reduction of their toxicity [32]. They were down-regulated in *A. alternata* in response to citral, including *CC77DRAFT_36175*, *CC77DRAFT_1015047*, and *CC77DRAFT_1026574*. Additionally, DEGs of oxidoreductase activity might be likewise vital to supply reducing power to protect from the damage of ROS accumulation, but they were significantly enriched in down-regulated DEGs. Moreover, sulfur metabolism, including sulfate assimilation and sulfate reduction, plays practical roles in stress tolerance [33,34]. Sulfur-containing defense compounds, including sulfide, glutathione, and various secondary metabolites, as well as sulfur-rich proteins, are crucial for fungal survival under abiotic stress [33,35]. The transcripts of eight DEGs in relation to sulfur metabolism were all repressed by citral in *A. alternata*. In connection with the results of T-AOC and catalase activity, fungal cells could not remove the accumulation of ROS in a timely manner after citral treatment. This might give rise to the sharp violation of ROS balance, and the disorder of oxidative stress eventually resulted in the disruption of cell structure and the disturbance of mycotoxin biosynthesis.

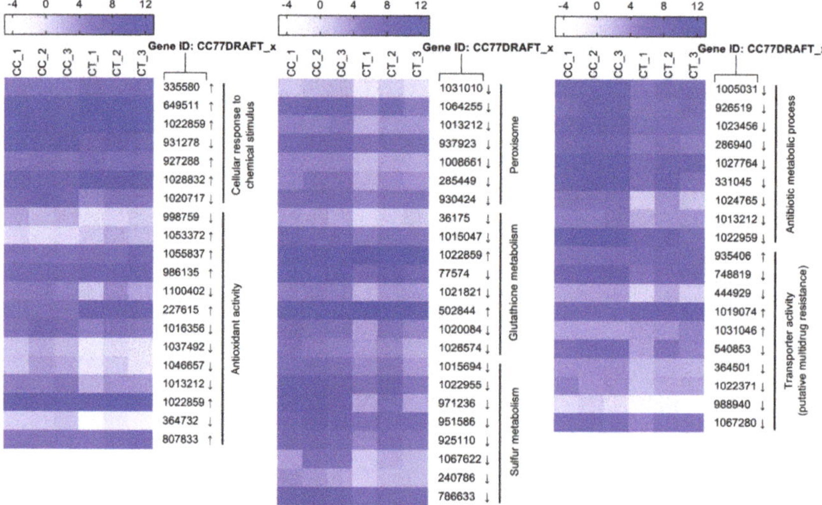

Figure 8. Comparative expression level of DEGs putatively responsible for stress response. Gene expression values are expressed as $\log_2$ TPM (transcripts per million reads). Arrows represent up-regulated (↑) and down-regulated (↓) DEGs.

There are some other defense systems that resist against abiotic stress, such as multidrug resistance [17]. Drug metabolic processes, including antibiotic metabolic processes, may assist in multidrug resistance. The results of GO enrichment demonstrated that these processes were partially impaired by citral. Furthermore, transporters, especially ATP-binding cassette (ABC) transporter and major facilitator superfamily (MFS) transporter, play an important role in the efflux capacity of xenobiotic compounds [36]. Correspondingly, putative multidrug resistance of transporter activity was notably inhibited by the essential oil, which was unfavorable to alleviate the stress in *A. alternata*.

2.6. Citral Interferes with the Expression of Genes Responsible for Mycotoxin Biosynthesis

The biosynthetic gene cluster responsible for AOH and AME biosynthesis has been elucidated [10]. It was sufficient for AOH formation by *pksI* (*CC77DRAFT_545549*) in *A. alternata* [10,11]. AME is the product of AOH methylation catalyzed by an *omtI* (*CC77DRAFT_1028551*) encoding methyl-transferase.

Another three enzymes (moxI, sdrI and doxI) were also involved in AOH modification. Additionally, a GAL4-like Zn(II)2Cys6 transcription factor expressed by *aohR* (*CC77DRAFT_1028550*) in the *pksI*-gene cluster modulated the transcriptional enhancement of the clustered synthase genes. The expression of four clustered genes (*pksI*, *omtI*, *sdrI* and *doxI*) showed down-regulation, as observed from RNA-Seq data, in the citral treatment (Figure 9A). To validate the analyses from RNA-Seq data, three clustered genes (*pksI*, *omtI* and *aohR*) directly involved in AOH and AME biosynthesis and regulation were chosen to be employed for quantitative reverse transcription PCR (qRT-PCR) analysis. The results of these gene expression patterns were similar to the outcomes of transcriptomic analysis (Figure 9A,B). This further demonstrated that citral could modulate the down-regulation of biosynthetic genes, including *pksI* and *omtI*.

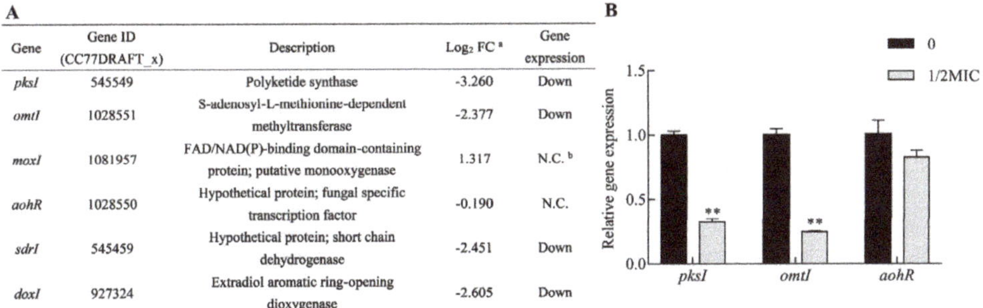

A

Gene	Gene ID (CC77DRAFT_x)	Description	Log₂ FC [a]	Gene expression
pksI	545549	Polyketide synthase	-3.260	Down
omtI	1028551	S-adenosyl-L-methionine-dependent methyltransferase	-2.377	Down
moxI	1081957	FAD/NAD(P)-binding domain-containing protein; putative monooxygenase	1.317	N.C. [b]
aohR	1028550	Hypothetical protein; fungal specific transcription factor	-0.190	N.C.
sdrI	545459	Hypothetical protein; short chain dehydrogenase	-2.451	Down
doxI	927324	Extradiol aromatic ring-opening dioxygenase	-2.605	Down

Figure 9. Transcriptional analysis of the clustered genes involved in AOH and AME production. (**A**) Table showing the results of gene expression from the analysis of RNA-Seq. [a] FC: fold change; [b] N.C.: no significant change in gene expression level. (**B**) Comparative analysis of gene expression by quantitative reverse transcription PCR (qRT-PCR). *PksI*, *omtI* and *aohR* were directly responsible for AOH and AME biosynthesis and regulation in *A. alternata*. The data are presented as the mean ± SEM ($n = 3$). MIC: minimum inhibitory concentration.

3. Discussion

A. alternata is an important phytopathogen, causing agricultural output losses and playing a tremendous role in food and feed safety due to the production of mycotoxins. Of these mycotoxins, AOH and AME were shown to be two of the most frequently contaminated mycotoxins [2,3]. To manage this contamination, essential oils have been shown to be environmentally friendly alternatives to common antifungal agents, especially in consideration of postharvest contamination of food and feed with *Alternaria* mycotoxins. Citral has been shown to be brilliant in suppressing fungal infection and mycotoxin contamination in *P. expansum* [20], *A. ochraceus* [37] and *Fusaria* [38]. Among them, the potential antifungal and antimycotoxigenic mechanisms were illustrated via transcriptomic profiling in the inhibition of *P. expansum* by the combination of cinnamaldehyde and citral [20]. Nevertheless, the distinction of these two essential oils and the exact action mode of citral alone was still not clearly known in resisting against *P. expansum*. Additionally, essential oil, to a great extent, played different roles in cellular response in various fungal species in previous studies [14]. Therefore, the inhibitory effects and mechanisms of citral suppressing the growth and mycotoxin production of *A. alternata* were uncovered in this study.

Citral exerted highly inhibitory effects on the mycelial growth and mycotoxin production of *A. alternata*. The antifungal efficiency was similar to that of cinnamaldehyde described by Xu et al. [25]. Citral has broad-spectrum antifungal activity, and it was much more effective than eugenol, geraniol, limonene and linalool, impairing *A. alternata* as well as *A. niger*, *A. flavus*, *F. moniliforme*, etc. [21]. Citral exhibited strong fungicide capacity by disrupting cell structure as shown by comparative microscope analysis with or without chemical stress. This antifungal finding can be regarded as similar to former reports on a number of essential oils such as cinnamaldehyde [25,39], eugenol [40], and thymol [41]. In addition, fungicidal activity was further demonstrated by the inhibition of spore germination. This result was consistent with that of *Nepeta rtanjensis* essential oil, which suppressed the elongation of the germ tube and spore germination, observed by light microscopy, as the concentration increased [42,43]. The sharp enhancement of fungal cell permeability was reflected by the abnormal and massive spillage of intracellular soluble proteins from macromolecular cytoplasmic components owing to cell damage by citral. In previous studies, intracellular reducing sugar and protein influenced the cell lysis rate by (*E*)-2-hexenal on *A. flavus* [44] or 7-demethoxytylophorine on *P. italicum* [45]. Accordingly, citral caused spore lysis in a dose-dependent manner from the result of released soluble protein in our study.

Fungal cell integrity is essential to adapt to the stress response caused by toxic compounds. The lipophilic terpenoid intrudes into plasma membranes and causes the disturbance of membrane integrity. Ergosterol is the major component of structural sterols, especially in the fungal cell membrane. The disorder of ergosterol may lead to much more fragility in response to stress, while the marked reduction of ergosterol could result from citral treatment in this work. The disruption of plasma membranes makes fungal cells so much more vulnerable as citral disturbs the exchange of substantial and energy metabolism via plasma membranes. This could be the mechanism of the cytotoxicity of citral against *A. alternata*. In combination with the result of scanning electron microscopy, exposure to the MIC of citral led to seriously deleterious damage to the cell plasma membrane. Furthermore, it would be necessary to carry out further studies to understand the actual relationship between the disruption of plasma membranes and the inhibition of mycotoxin production. Moreover, fatty acid biosynthesis also influences the membrane fluidity and rigidity. Correspondingly, the expression pattern of DEGs involved in the fatty acid biosynthesis pathway was significantly impaired by citral. In a previous study, the other monoterpene, *d*-limonene, induced cytotoxicity on the alteration of the cell wall as the main target but not on the plasma membrane in *Saccharomyces cerevisiae* [46]. However, another monoterpene, α-terpinene, displayed strong antagonistic activity against *S. cerevisiae* and the overexpression of numerous genes in relation to the cell wall and also membrane biogenesis [47]. In our study, the considerable alteration of the cell surface and membrane was eventually revealed by the results of ergosterol content, shrunken and wrinkled cell surfaces and RNA-Seq data under the citral stress in this report.

Fungal oxidative damage has been considered as an essential factor for the antifungal and antimycotoxigenic properties of essential oils [20,28,29]. ROS are inevitably produced by essential oils in response to stress, which can react with intracellular components and lead to chemical damage, such as lipid peroxidation, protein oxidation, etc. Overaccumulation of ROS can eventually cause defective cells and even lethal cells. On the other hand, fungal antioxidative defense systems of ROS stress have been evolved to detoxify ROS, including enzymatic systems, such as catalases and superoxide dismutases, and also non-enzymatic systems, such as glutathione and thioredoxin [32]. ROS exert multiple roles as signaling molecules for fungal growth, development, biosynthesis of secondary metabolites and other cell processes. Consequently, the maintenance of ROS level is critical to the survival of the fungal life cycle. In this work, the total antioxidative capacity of *A. alternata* exposed to citral was markedly reduced. Correspondingly, ROS could accumulate and this could be detrimental to fungal survival under the citral condition. In addition, catalase activity was also demonstrated to be lower after citral treatment through an enzyme activity assay and RNA-Seq data. This would imply that hydrogen peroxide, the substrate of catalase, could be difficult to scavenge in time. These results demonstrated that the balance of ROS level was disturbed after the exposure to citral. In recent studies, similar observations revealed that the antifungal and antimycotoxigenic characteristics of essential oils were highly connected with the induction of ROS formation [28,29,48]. For instance, the inhibitory effects of cinnamaldehyde against mycelia growth and aflatoxin B_1 could be attributed to the perturbation of redox status [28]. Enriched analysis of DEGs also demonstrated that oxidoreductase activity, glutathione metabolism and sulfur metabolism were overall down-regulated in *A. alternata* in response to citral stress. These cellular processes were closely related to ROS maintenance. Nevertheless, the abnormality of these cellular processes was unfavorable to the survival of *A. alternata* exposed to citral. Interestingly, siderophore was demonstrated to be resistant against oxidative stress [49]. A nonribosomal peptide synthetase encoding gene (*AaNPS6*), involved in siderophore biosynthesis [50], was highly regulated after citral treatment. However, the increase seemed to be inadequate to alleviate the accumulation of oxidative stress and fungal damage in this study, as evidenced by shrunken and wrinkled cell surfaces and the reduction of total antioxidative capacity. On the other hand, the alteration of carbon flow on the biosynthesis of secondary metabolites might be the reason for the marked reduction of AOH and AME. In addition, multidrug resistance facilitates fungal defense against xenobiotic stress. Except for potential drug metabolic processes, fungal ABC and MFS transporter proteins are essential for the efflux of toxic compounds for antifungal drug resistance [36,51]. A number of these functional genes in association with cellular multidrug resistance were down-regulated by citral. This reduction tended to be harmful to fungal survival with exposure to citral.

In this study, it was demonstrated that the repression of biomass was not the only cause of the reduction of AOH and AME production. This could be likewise attributed to the obstacle of mycotoxin biosynthetic genes under the citral condition. *PksI* and *omtI*, involved in mycotoxin biosynthesis, were less transcribed, while the expression of *aohR* encoding a specific transcriptional factor showed a slight decrease but no significant change under citral stress. *PksI* was sufficient for AOH formation in *A. alternata*, which was further confirmed by heterougous expression in *A. oryzae* [10]. In addition, *omtI* encoded a methyl-transferase, catalyzing the transformation from AOH to AME. Mycotoxin production could be significantly reduced by down-regulating the transcriptional level of the two enzymatic genes in this work. Similar results of gene expression were observed in the mycotoxin biosynthetic pathway after essential oil treatment. *AcOTApks* and *acOTAnrps*, directly responsible for ochratoxin A (OTA) biosynthesis, were obviously down-regulated by essential oils such as fennel and cardamom, even though they did not affect the growth of *A. carbonarius* S402 [52]. However, a recent study demonstrated that both the mycelial growth and transcriptional level of *pks* and *nrps* involved in OTA production were down-regulated in *A. ochraceus* fc-1 in response to cinnamaldehyde [39]. This reveals that there exist different inhibitory mechanisms of essential oils suppressing fungal growth and mycotoxin production. Another similar study was conducted, in which the expression of aflatoxin biosynthetic genes, such as *aflD*, *aflM*, *aflO*, *aflP*, and *aflQ*, was suppressed by turmeric essential oil [53]. Interestingly, one transcriptional regulator gene, *aflR*, in the aflatoxin biosynthetic gene cluster was significantly inhibited, while the other regulation enhancer, *aflS*, was observed to have no obvious changes, like *aohR*, in terms of transcriptional level. The expressions of *nor-1* (*aflD*), *ver-1* (*aflM*) and *omt-A* (*aflP*) were likewise repressed in toxigenic *A. parasiticus* exposed to *Zataria multiflora Boiss* essential oil [54,55].

In conclusion, the antifungal and antimycotoxigenic mechanisms of citral on the growth and mycotoxin production of *A. alternata* were unraveled in this study. Citral caused irreversible damage to the spore ultrastructure. In addition, citral was able to disturb oxidative balance, disrupt cell integrity, repress transporter activity and down-regulate biosynthetic genes of AOH and AME, including *pksI* and *omtI*. Additionally, citral, as well as other essential oils like cinnamaldehyde, eugenol, thymol, etc. has been generally recognized as safe and registered for food flavoring in the European Union, regarded as having no toxicity and preservative potential [15,56]. However, citral was also indicated as a contact irritant and a contact allergen. Therefore, it is more applicable for unprocessed food raw materials, such as cereal storage. Besides, there is the great challenge of its strong fragrance even at relatively low concentrations. Encapsulation could potentially provide a novel strategy for its usage to minimize the organoleptic impact, especially nanoencapsulation. In general, in view of its environmental friendliness and highly fungicidal characteristics, citral may be a potentially promising alternative as a chemical fungicide for cereal storage.

4. Materials and Methods

4.1. Chemicals, Strain and Culture Conditions

Mycotoxin standards of AOH and AME were purchased from Romer Labs (Newark, DE, USA). Ergosterol was obtained from Sigma-Aldrich (St. Louis, MO, USA). Citral was obtained from Jiangxi Xuesong Natural Medicinal Oil Co. LTD (Jiangxi, China).

A. alternata ATCC 66981 was acquired from American Type Culture Collection and cultured on Potato Dextrose Agar (PDA) for 6 days at 25 °C. The spores were washed with 0.1% (*v/v*) Tween-80 and adjusted to 1×10^5 spores/mL using a haemocytometer for the following culture.

4.2. Antifungal Effects of Citral on Mycelial Growth and Spore Germination

The inhibitory effects of citral were determined by the serial twice dilution method in Potato Dextrose Broth (PDB) medium. The citral was dissolved in ethanol, and then the solution was saved as the stock. The stock solution was diluted to the final concentrations of 0, 0.0625, 0.125, 0.25, 0.5 and 1.0 μL/mL (0, 55.625, 111.25, 222.5, 445 and 890 μg/mL). Quantities of 1 mL of the *A. alternata* spores were inoculated into the medium with the serial concentrations of citral. The strain was then cultured for 6 days at 25 °C, 180 r/min for mycelial weight measurement. The MIC was regarded as the minimum concentration with no mycelium growth of *A. alternata*. An amount of 200 μL of the culture with no mycelia growth was spread onto citral-free PDA and cultured for 6 days at 25 °C. The MFC was determined as the minimum concentration without any mycelium growth on PDA. The fermented mycelium was collected and washed with sterile water to completely remove the medium residues. The mycelia were dried individually by vacuum freeze drying and then weighed.

The trials of fungal spore germination were performed according to a previous method [42,53] with minor modification. The spores were collected and resuspended in sterile double distilled water, and the concentration was adjusted to 1×10^5 spores/mL. One-milliliter aliquots of the spores were mixed with the liquid PDA medium at 45 °C incubation, where citral was added to the final concentrations as described above. The mixture was spread quickly and solidified on Petri dishes. Subsequently, the spores were incubated in the dark at 25 °C. After 24 h of culture, the germinating spores were terminated and dyed with lactophenol cotton blue. The spore germination was judged by the rule that the length of the germ tube must be at least half of the spore diameter, as described by Grbić et al. [42]. At least 200 spores, whether germinated or not, were randomly counted and estimated. Finally, the rates of spore germination were individually calculated.

4.3. Determination of Mycotoxin Production

The supernatant of the fermented fungus treated with 0, 1/4MIC and 1/2MIC of citral was separately collected and taken for the mycotoxin measurement. Four milliliters of the acetonitrile were added into 1 mL of the supernatant and mixed thoroughly with a vortex shaker. The mixture was evaporated to dryness by a gentle nitrogen stream at 50 °C. The residue was resolved with 1 mL of acetonitrile–water (30:70, *v/v*) and then filtered into a vial with polytetrafluoroethylene (PTFE) membrane. From the vial, 20 μL was injected and assayed by a HPLC-UV/FLD system (Agilent Technologies, Santa Clara, CA, USA) comprising a UV/Visible Detector (258 nm) and FLD Detector (excitation wavelength, 328 nm; emission wavelength, 405 nm), using a reversed phase TC-C18 column (5 μm, 4.6 mm × 250 mm, Agilent, Santa Clara, CA, USA). The column temperature was set at 35 °C, and the flow rate was 1.5 mL/min. The mobile phase was double distilled water and acetonitrile containing 1 mM of oxalic acid. The mycotoxins, AOH and AME, were separated by the mobile phase with a gradient elution rising linearly from 20% to 70% acetonitrile for 15 min at the beginning, maintaining 70% acetonitrile for 1 min, then decreasing to the former 20% acetonitrile for 1.5 min, and finally maintaining 20% acetonitrile for 4.5 min. The mycotoxin concentrations were finally confirmed with UPLC-MS (TQ-S, Waters Micromass, Manchester, UK) under the guidance of Wang et al. [57].

4.4. Transcriptome Analysis

The spore suspensions of *A. alternata* (1×10^5 spores/mL) were separately inoculated in PDB with or without 1/2MIC of citral. They were cultured in a shaker incubator at 25 °C, 180 r/min for 6 days. Mycelia were subsequently collected for the transcriptome analysis. RNA extraction, cDNA library construction, RNA-Seq and the following analysis was performed by ShangHai Majorbio Bio-pharm Technology Co., Ltd. (Shanghai, China). Total RNA was extracted by TRIzol reagent (Invitrogen, Life Technologies, Carlsbad, CA, USA) [58]. The mRNA of each sample was enriched by oligo (dT) magnetic beads and then fragmented into the fragmentation buffer. First-strand cDNA was obtained by reverse transcriptase using random hexamers, and then both cDNA strands were synthesized. The purity and concentration of each RNA sample were detected by a NanoDrop 2000 (Thermo Fisher Scientific, Waltham, MA, USA). The integrity of RNA was calculated by 1% agarose gel electrophoresis, and the value of RNA Integrity Number (RIN) was determined by an Agilent 2100 Bioanalyzer using an RNA 6000 Nano kit (Agilent Technologies, Santa Clara, CA, USA). The cDNA library was constructed by the Illumina TruseqTM RNA Sample Preparation kit and sequenced (2×150 bp read length) by Illumina HiSeq4000 platforms (San Diego, CA, USA).

The clean data were obtained by removing adaptor sequences, low-quality reads, sequences with 10% higher N rate (N: ambiguous bases information), and over-short sequences of the length from the raw sequenced reads. The clean reads were mapped to the reference genome of *A. alternata* SRC1lrK2f(GCA_001642055) (http://fungi.ensembl.org/Alternaria_alternata_gca_001642055/Info/Index) by the software of Hisat2 [59]. The mapped reads were assembled by StringTie [60,61].

The sequence annotation was analyzed by DIAMOND software [62], searching against the NCBI non-redundant (NR) protein database, Swiss-Prot, and EggNOG database. The classification of GO terms was carried out by BLAST2GO [63]. The protein family was annotated by HMMER3 [64], searching the hidden Markov model (HMM) of the established protein domain against the Pfam database. KEGG pathway annotation was performed by KOBAS 2.1 [65] against the KEGG database.

The read counts were quantified by RSEM software in terms of transcripts per million reads (TPM) [66]. The DEGs were analyzed by DESeq2 [67] and considered as statistical significance of gene expression differences at FDR < 0.05 and the absolute value of $\mathrm{Log_2}$ FC ≥1. GO enrichment was analyzed in terms of DEGs by Goatools at FDR < 0.05 [68]. The KEGG pathway enrichment analysis of the transcript was conducted by KOBAS 2.1 [65] with respect to DEGs by FDR < 0.05.

4.5. Detection of Fungal Ergosterol Content

The fungal ergosterol was extracted and quantified by the minor modified method, under the reference described by Bomfi et al. [26]. Each fungal sample treated with 0 or 1/2MIC of citral was separately homogenized with a glass pestle and completely washed into the solution containing 20 mL of methanol, 5 mL of ethanol and 2.0 g KOH. This was rotated at 25 °C, 250 r/min for 20 min and then incubated for 40 min at 70 °C. Then, this solution was cooled and mixed with 5 mL deionized water. Two milliliters of the supernatant were collected by centrifuging at 10,000 r/min for 10 min. The ergosterol was extracted using an equal volume of n-hexane. The organic phase was evaporated at 50 °C using nitrogen flushing and resolved in methanol. The ergosterol was detected at 282 nm, at the flow rate of 1.5 mL/min, for 15 min using 100% acetonitrile as a mobile phase by a 1260 infinity HPLC-UV system (Agilent Technologies, Santa Clara, CA, USA) after filtration with 0.2 μm polytetrafluoroethylene (PTFE) membrane. The injection volume was 50 μL, and the HPLC separation was performed with a TC-C18 column with a 5 μm particle size (4.6 mm × 250 mm, Agilent, Santa Clara, CA, USA). The ergosterol content was expressed by dividing the fungal mycelia weight of *A. alternata*.

4.6. Release of Intracellular Protein

The release of the intracellular soluble protein from *A. alternata* spores was assayed according to the method of Chen et al. [45] with some modification. The spores of *A. alternata* were collected from 6 days culture and washed thrice with sterile double distilled water. The concentration of the released protein was determined after removing the fungal spores and cell debris of *A. alternata*. The spores were centrifuged and then resuspended at the final concentration of 1×10^5 spores/mL in 0.01 M of sterile phosphate buffered solution (PBS, pH 7.2–7.4, Solarbio Life Sciences Inc., Beijing, China) containing different concentrations of citral (0, 1/4MIC, 1/2MIC, MIC and MFC) for 24 h. The leaked protein was determined by bicinchoninic acid (BCA) reagent using a BCA Protein Assay Kit (Solarbio, Beijing, China) and quantified by an EnVision Multilabel Reader (PerkinElmer, Boston, MA, USA) using Bovine Serum Albumin (BSA) as the protein standard.

4.7. Scanning Electron Microscopy Analysis

For the microscopic morphology observation, the spores were treated in the medium with or without the MIC of citral. The spores were collected and washed twice with 0.1 M of PBS. Subsequently, they were fixed in 2.5% glutaraldehyde overnight. On the second day, they were centrifuged and washed twice with 0.1 M of PBS. The samples were dehydrated by serial ethanol solutions (30%, 50%, 70%, 80%, 90%) for a period of 15 min. Then, the samples were washed twice in absolute ethanol for 20 min and placed in tertiary butanol twice to completely replace ethanol for 30 min. They were dried through vacuum freeze drying. The samples were mounted on a stub and coated with gold. Finally, the micromorphology observation was individually calculated by Hitachi S-3400 scanning electron microscope at 5 kV accelerating voltage (Tokyo, Japan).

4.8. Analysis of Total Antioxidant Capacity and Catalase Activity

The mycelia of *A. alternata* treated with diverse concentrations of citral (0 or 1/2MIC) were collected and washed three times with sterile distilled water. The samples were dried in a vacuum and ground thoroughly using liquid nitrogen. They were separately resuspended in equal volumes of 0.01 M PBS. They were centrifuged, and the supernatants were used for the determination of total antioxidant capacity and catalase activity. The comparison of total antioxidant capacity and catalase activity was calibrated by the protein concentration of each sample. The protein concentration was assayed by a BCA Protein Assay Kit (Solarbio, Beijing, China). All the assays were conducted on an EnVision Multilabel Reader (PerkinElmer, Boston, MA, USA). All the measurements were carried out in triplicate.

Total antioxidant capacity was determined by the T-AOC Assay Kit (Beyotime, Shanghai, China) with the FRAP method. The standard curve was calculated using various concentrations of $FeSO_4$ (0, 0.1, 0.25, 0.5, 1.0, 2.5, 5.0 mM). For the FRAP method, the total antioxidant capacity was expressed by the concentration of $FeSO_4$ standard solution.

Catalase activity was analyzed by a catalase Assay Kit (Beyotime, Shanghai, China). The supplied H_2O_2 in the kit was diluted 100 times, and then the absorbance was detected at 240 nm. The actual concentration (mM) was calculated by the formula of $22.94 \times A_{240nm}$ following the instructions of the kit. The standard curve was measured using the actual various concentrations of H_2O_2 (0, 0.625, 1.25, 2.5, 3.75, 5.0 mM). One unit of enzyme activity (1 U) was determined to catalyze the decomposition of 1 μmol of H_2O_2 in 1 minute at 25 °C, pH 7.0.

4.9. Transcriptional Validation of Biosynthetic Genes Involved in Mycotoxin Production

Total RNA was separately extracted from the treated samples for the following qRT-PCR analysis by an EasyPure Plant RNA Kit (TransGen Biotech, Beijing, China). During the process, the DNA residue was digested by RNase-free DNase I. Total RNA was quantified using a Merinton SMA4000 UV-VIS Spectrophotometer (Ann Arbor, MI, U.S.A) and equally adjusted. The cDNA template was separately synthesized by reverse transcription using the kit of TransScript One-Step gDNA Removal and cDNA Synthesis SuperMix (TransGen Biotech Inc., Beijing, China). The 20 μL quantity of reaction mixture included 10 μL of 2× SYBR Green Master Mix (Applied Biosystems, Foster City, CA, USA), 8.4 μL of double distilled water, 0.6 μL of primer pair (each primer, 10 μM), and 1.0 μL of cDNA as a template. The primers were designed by Primer Premier 6 software based on the sequences of gene transcripts in *A. alternata* (Table S3). The qRT-PCR was then performed by StepOne Plus Real-time PCR systems (Applied Biosystems, Foster City, CA, USA). The relative transcriptional level was separately determined by the $2^{-\Delta\Delta t}$ method.

4.10. Statistical Analyses

All the results were calculated as the mean ± SEM for at least triplicates. The mean differences of the data were compared by analysis of variance (ANOVA) using Tukey's post hoc test, following the significance at $p < 0.05$ by IBM SPSS statistics 23.0 (IBM Inc., Armonk, NY, USA). The corresponding figures were processed by GraphPad Prism 7.0 (GraphPad Software Inc., San Diego, CA, USA).

Supplementary Materials: The following are available online at http://www.mdpi.com/2072-6651/11/10/553/s1, Figure S1: Ergosterol content of *A. alternata* in response to different concentrations of citral, Table S1: Statistics of RNA-Seq data from *A. alternata*, Table S2: Identification and functional analysis of gene expression in *A. alternata*, Table S3: Primer sequences designed for quantitative reverse transcription PCR (qRT-PCR) in *A. alternata*.

Author Contributions: Conceptualization, L.W. and M.W.; methodology, L.W.; software, L.W.; validation, L.W. and M.W.; formal analysis, L.W.; investigation, L.W., N.J. and D.W.; resources, L.W., N.J. and D.W.; data curation, L.W. and M.W.; writing-original draft preparation, L.W.; writing-review and editing, L.W. and M.W.; visualization, L.W.; project administration, M.W.; funding acquisition, L.W. and M.W.

Funding: This work was supported by the Beijing Natural Science Foundation (Grant No. 6184038), the Open Project of the Laboratory of Quality & Safety Risk Assessment for Agro-products (Beijing), Ministry of Agriculture and Rural Affairs (Grant No. KFRA201801) and the Innovation and Capacity-building Projects by Beijing Academy of Agriculture and Forestry Sciences (Grant No. KJCX20180408).

Conflicts of Interest: The authors declare no conflict of interest.

References

1. EFSA on Contaminants in the Food Chain (CONTAM). Scientific Opinion on the risks for animal and public health related to the presence of *Alternaria* toxins in feed and food. *EFSA J.* **2011**, *9*, 2407. [CrossRef]
2. Ostry, V. *Alternaria* mycotoxins: An overview of chemical characterization, producers, toxicity, analysis and occurrence in foodstuffs. *World Mycotoxin J.* **2008**, *1*, 175–188. [CrossRef]
3. Logrieco, A.; Moretti, A.; Solfrizzo, M. *Alternaria* toxins and plant diseases: An overview of origin, occurrence and risks. *World Mycotoxin J.* **2009**, *2*, 129–140. [CrossRef]
4. Lee, H.B.; Patriarca, A.; Magan, N. *Alternaria* in food: Ecophysiology, mycotoxin production and toxicology. *Mycobiology* **2015**, *43*, 93–106. [CrossRef] [PubMed]
5. Pfeiffer, E.; Eschbach, S.; Metzler, M. *Alternaria* toxins: DNA strand-breaking activity in mammalian cells *in vitro*. *Mycotoxin Res.* **2007**, *23*, 152–157. [CrossRef] [PubMed]
6. Liu, G.; Qian, Y.; Zhang, P.; Dong, Z.; Shi, Z.; Zhen, Y.; Miao, J.; Xu, Y. Relationships between *Alternaria alternata* and oesophageal cancer. *IARC Sci. Publ.* **1991**, *105*, 258–262.
7. Liu, G.T.; Qian, Y.Z.; Zhang, P.; Dong, W.H.; Qi, Y.M.; Guo, H.T. Etiological role of *Alternaria alternata* in human esophageal cancer. *Chin. Med. J.* **1992**, *105*, 394–400. [PubMed]
8. Sáenz-de-Santamaría, M.; Postigo, I.; Gutierrez-Rodríguez, A.; Cardona, G.; Guisantes, J.A.; Asturias, J.; Martínez, J. The major allergen of *Alternaria alternata* (Alt a 1) is expressed in other members of the Pleosporaceae family. *Mycoses* **2006**, *49*, 91–95. [CrossRef] [PubMed]
9. Hayes, T.; Rumore, A.; Howard, B.; He, X.; Luo, M.; Wuenschmann, S.; Chapman, M.; Kale, S.; Li, L.; Kita, H.; et al. Innate immunity induced by the major allergen Alt a 1 from the fungus *Alternaria* is dependent upon toll-like receptors 2/4 in human lung epithelial cells. *Front. Immunol.* **2018**, *9*, 1507. [CrossRef] [PubMed]
10. Wenderoth, M.; Garganese, F.; Schmidt-Heydt, M.; Soukup, S.T.; Ippolito, A.; Sanzani, S.M.; Fischer, R. Alternariol as virulence and colonization factor of *Alternaria alternata* during plant infection. *Mol. Microbiol.* **2019**. [CrossRef] [PubMed]
11. Chooi, Y.-H.; Muria-Gonzalez, M.J.; Mead, O.L.; Solomon, P.S. *SnPKS19* encodes the polyketide synthase for alternariol mycotoxin biosynthesis in the wheat pathogen *Parastagonospora nodorum*. *Appl. Environ. Microbiol.* **2015**, *81*, 5309–5317. [CrossRef] [PubMed]

12. Demuner, A.J.; Barbosa, L.C.A.; Miranda, A.C.M.; Geraldo, G.C.; Da Silva, C.M.; Giberti, S.; Bertazzini, M.; Forlani, G. The fungal phytotoxin alternariol 9-methyl ether and some of its synthetic analogues inhibit the photosynthetic electron transport chain. *J. Nat. Prod.* **2013**, *76*, 2234–2245. [CrossRef] [PubMed]

13. Dwivedy, A.K.; Kumar, M.; Upadhyay, N.; Prakash, B.; Dubey, N.K. Plant essential oils against food borne fungi and mycotoxins. *Curr. Opin. Food Sci.* **2016**, *11*, 16–21. [CrossRef]

14. Bakkali, F.; Averbeck, S.; Averbeck, D.; Idaomar, M. Biological effects of essential oils—A review. *Food Chem. Toxicol.* **2008**, *46*, 446–475. [CrossRef] [PubMed]

15. Hyldgaard, M.; Mygind, T.; Meyer, R.L. Essential oils in food preservation: Mode of action, synergies, and interactions with food matrix components. *Front. Microbiol.* **2012**, *3*, 12. [CrossRef] [PubMed]

16. Tomazoni, E.Z.; Pansera, M.R.; Pauletti, G.F.; Moura, S.; Ribeiro, R.T.S.; Schwambach, J. In vitro antifungal activity of four chemotypes of *Lippia alba* (Verbenaceae) essential oils against *Alternaria solani* (Pleosporeaceae) isolates. *An. Acad. Bras. Cienc.* **2016**, *88*, 999–1010. [CrossRef]

17. OuYang, Q.; Tao, N.; Jing, G. Transcriptional profiling analysis of *Penicillium digitatum*, the causal agent of citrus green mold, unravels an inhibited ergosterol biosynthesis pathway in response to citral. *BMC Genom.* **2016**, *17*, 1–16. [CrossRef] [PubMed]

18. Zheng, S.; Jing, G.; Wang, X.; Ouyang, Q.; Jia, L.; Tao, N. Citral exerts its antifungal activity against *Penicillium digitatum* by affecting the mitochondrial morphology and function. *Food Chem.* **2015**, *178*, 76–81. [CrossRef]

19. Tao, N.; OuYang, Q.; Jia, L. Citral inhibits mycelial growth of *Penicillium italicum* by a membrane damage mechanism. *Food Control* **2014**, *41*, 116–121. [CrossRef]

20. Wang, Y.; Feng, K.; Yang, H.; Zhang, Z.; Yuan, Y.; Yue, T. Effect of cinnamaldehyde and citral combination on transcriptional profile, growth, oxidative damage and patulin biosynthesis of *Penicillium expansum*. *Front. Microbiol.* **2018**, *9*, 597. [CrossRef]

21. Kishore, G.K.; Pande, S.; Harish, S. Evaluation of essential oils and their components for broad-spectrum antifungal activity and control of late leaf spot and crown rot diseases in peanut. *Plant Dis.* **2007**, *91*, 375–379. [CrossRef]

22. Liang, D.; Xing, F.; Selvaraj, J.N.; Liu, X.; Wang, L.; Hua, H.; Zhou, L.; Zhao, Y.; Wang, Y.; Liu, Y. Inhibitory effect of cinnamaldehyde, citral, and eugenol on aflatoxin biosynthetic gene expression and aflatoxin B_1 biosynthesis in *Aspergillus flavus*. *J. Food Sci.* **2015**, *80*, M2917–M2924. [CrossRef]

23. Mahmoud, A.-L.E. Antifungal action and antiaflatoxigenic properties of some essential oil constituents. *Lett. Appl. Microbiol.* **1994**, *19*, 110–113. [CrossRef]

24. Malavazi, I.; Goldmanm, G.H.; Brown, N.A. The importance of connections between the cell wall integrity pathway and the unfolded protein response in filamentous fungi. *Brief. Funct. Genomics* **2014**, *13*, 456–470. [CrossRef]

25. Xu, L.; Tao, N.; Yang, W.; Jing, G. Cinnamaldehyde damaged the cell membrane of *Alternaria alternata* and induced the degradation of mycotoxins *in vivo*. *Ind. Crops Prod.* **2018**, *112*, 427–433. [CrossRef]

26. Bomfi, N.S.; Nakassugi, L.P.; Oliveira, J.F.P.; Kohiyama, C.Y.; Mossini, S.A.G.; Grespan, R.; Nerilo, S.B.; Mallmann, C.A.; Filho, B.A.A.; Machinski, M.J. Antifungal activity and inhibition of fumonisin production by *Rosmarinus officinalis* L. essential oil in *Fusarium verticillioides* (Sacc.) Nirenberg. *Food Chem.* **2015**, *166*, 330–336.

27. Abhishek, R.U.; Thippeswamy, S.; Manjunath, K.; Mohana, D.C. Antifungal and antimycotoxigenic potency of *Solanum torvum* Swartz. leaf extract: Isolation and identification of compound active against mycotoxigenic strains of *Aspergillus flavus* and *Fusarium verticillioides*. *J. Appl. Microbiol.* **2015**, *119*, 1624–1636. [CrossRef]

28. Sun, Q.; Shang, B.; Wang, L.; Lu, Z.; Liu, Y. Cinnamaldehyde inhibits fungal growth and aflatoxin B_1 biosynthesis by modulating the oxidative stress response of *Aspergillus flavus*. *Appl. Microbiol. Biotechnol.* **2016**, *100*, 1355–1364. [CrossRef]

29. Kumar, K.N.; Venkataramana, M.; Allen, J.A.; Chandranayaka, S.; Murali, H.S.; Batra, H.V. Role of *Curcuma longa* L. essential oil in controlling the growth and zearalenone production of *Fusarium graminearum*. *LWT-Food Sci. Technol.* **2016**, *69*, 522–528. [CrossRef]

30. Nordgren, M.; Fransen, M. Peroxisomal metabolism and oxidative stress. *Biochimie* **2014**, *98*, 56–62. [CrossRef]

31. Titorenko, V.I.; Rachubinski, R.A. The life cycle of the peroxisome. *Nat. Rev. Mol. Cell Biol.* **2001**, *2*, 357–368. [CrossRef]

32. Montibus, M.; Pinson-Gadais, L.; Richard-Forget, F.; Barreau, C.; Ponts, N. Coupling of transcriptional response to oxidative stress and secondary metabolism regulation in filamentous fungi. *Crit. Rev. Microbiol.* **2015**, *41*, 295–308. [CrossRef]

33. Sieńko, M.; Natorff, R.; Skoneczny, M.; Kruszewska, J.; Paszewski, A.; Brzywczy, J. Regulatory mutations affecting sulfur metabolism induce environmental stress response in *Aspergillus nidulans*. *Fungal Genet. Biol.* **2014**, *65*, 37–47. [CrossRef]

34. Rausch, T.; Wachter, A. Sulfur metabolism: A versatile platform for launching defence operations. *Trends Plant Sci.* **2005**, *10*, 503–509. [CrossRef]

35. Gremel, G.; Dorrer, M.; Schmoll, M. Sulphur metabolism and cellulase gene expression are connected processes in the filamentous fungus *Hypocrea jecorina* (anamorph *Trichoderma reesei*). *BMC Microbiol.* **2008**, *8*, 174. [CrossRef]

36. Morschhäuser, J. Regulation of multidrug resistance in pathogenic fungi. *Fungal Genet. Biol.* **2010**, *47*, 94–106. [CrossRef]

37. Hua, H.; Xing, F.; Selvaraj, J.N.; Wang, Y.; Zhao, Y.; Zhou, L.; Liu, X.; Liu, Y. Inhibitory effect of essential oils on *Aspergillus ochraceus* growth and ochratoxin A production. *PLoS ONE* **2014**, *9*, e108285. [CrossRef]

38. Morcia, C.; Tumino, G.; Ghizzoni, R.; Bara, A.; Salhi, N.; Terzi, V. In vitro evaluation of sub-lethal concentrations of plant-derived antifungal compounds on *Fusaria* growth and mycotoxin production. *Molecules* **2017**, *22*, 1271. [CrossRef]

39. Wang, L.; Jin, J.; Liu, X.; Wang, Y.; Liu, Y.; Zhao, Y.; Xing, F. Effect of cinnamaldehyde on morphological alterations of *Aspergillus ochraceus* and expression of key genes involved in ochratoxin A biosynthesis. *Toxins* **2018**, *10*, 340. [CrossRef]

40. Latifah-Munirah, B.; Himratul-Aznita, W.H.; Mohd Zain, N. Eugenol, an essential oil of clove, causes disruption to the cell wall of *Candida albicans* (ATCC 14053). *Front. Life Sci.* **2015**, *8*, 231–240. [CrossRef]

41. Morcia, C.; Malnati, M.; Terzi, V. In vitro antifungal activity of terpinen-4-ol, eugenol, carvone, 1,8-cineole (eucalyptol) and thymol against mycotoxigenic plant pathogens. *Food Addit. Contam. Part A* **2012**, *29*, 415–422.

42. Grbić, M.L.; Stupar, M.; Vukojević, J.; Grubišić, D. Inhibitory effect of essential oil from *Nepeta rtanjensis* on fungal spore germination. *Cent. Eur. J. Biol.* **2011**, *6*, 583–586. [CrossRef]

43. Grbić, M.L.; Stupar, M.; Vukojević, J.; Soković, M.; Mišić, D.; Grubišić, D.; Ristić, M. Antifungal activity of *Nepeta rtanjensis* essential oil. *J. Serbian Chem. Soc.* **2008**, *73*, 961–965. [CrossRef]

44. Ma, W.; Zhao, L.; Zhao, W.; Xie, Y. (*E*)-2-Hexenal, as a potential natural antifungal compound, inhibits *Aspergillus flavus* spore germination by disrupting mitochondrial energy metabolism. *J. Agric. Food Chem.* **2019**, *67*, 1138–1145. [CrossRef]

45. Chen, C.; Qi, W.; Peng, X.; Chen, J.; Wan, C. Inhibitory effect of 7-demethoxytylophorine on *Penicillium italicum* and its possible mechanism. *Microorganisms* **2019**, *7*, 36. [CrossRef]

46. Brennan, T.C.R.; Krömer, J.O.; Nielsen, L.K. Physiological and transcriptional responses of *Saccharomyces cerevisiae* to d-limonene show changes to the cell wall but not to the plasma membrane. *Appl. Environ. Microbiol.* **2013**, *79*, 3590–3600. [CrossRef]

47. Parveen, M.; Hasan, M.K.; Takahashi, J.; Murata, Y.; Kitagawa, E.; Kodama, O.; Iwahashi, H. Response of *Saccharomyces cerevisiae* to a monoterpene: Evaluation of antifungal potential by DNA microarray analysis. *J. Antimicrob. Chemother.* **2004**, *54*, 46–55. [CrossRef]

48. Tian, J.; Ban, X.; Zeng, H.; He, J.; Chen, Y.; Wang, Y. The mechanism of antifungal action of essential oil from dill (*Anethum graveolens* L.) on *Aspergillus flavus*. *PLoS ONE* **2012**, *7*, e30147. [CrossRef]

49. Chen, L.H.; Yang, S.L.; Chung, K.R. Resistance to oxidative stress via regulating siderophore-mediated iron acquisition by the citrus fungal pathogen *Alternaria alternata*. *Microbiology* **2014**, *160*, 970–979. [CrossRef]

50. Chen, L.H.; Lin, C.H.; Chung, K.R. A nonribosomal peptide synthetase mediates siderophore production and virulence in the citrus fungal pathogen *Alternaria alternata*. *Mol. Plant Pathol.* **2013**, *14*, 497–505. [CrossRef]

51. Paul, S.; Moye-Rowley, W.S. Multidrug resistance in fungi: Regulation of transporter-encoding gene expression. *Front. Physiol.* **2014**, *5*, 143. [CrossRef] [PubMed]

52. El Khour, R.; Atoui, A.; Verheecke, C.; Maroun, R.; El Khoury, A.; Mathieu, F. Essential oils modulate gene expression and ochratoxin A production in *Aspergillus carbonarius*. *Toxins* **2016**, *8*, 242. [CrossRef] [PubMed]

53. Hu, Y.; Zhang, J.; Kong, W.; Zhao, G.; Yang, M. Mechanisms of antifungal and anti-aflatoxigenic properties of essential oil derived from turmeric (*Curcuma longa* L.) on *Aspergillus flavus*. *Food Chem.* **2017**, *220*, 1–8. [CrossRef] [PubMed]

54. Yahyaraeyat, R.; Khosravi, A.R.; Shahbazzadeh, D.; Khalaj, V. The potential effects of *Zataria multiflora Boiss* essential oil on growth, aflatoxin production and transcription of aflatoxin biosynthesis pathway genes of toxigenic *Aspergillus parasiticus*. *Braz. J. Microbiol.* **2013**, *44*, 649–655. [CrossRef] [PubMed]

55. Cleveland, T.E.; Yu, J.; Fedorova, N.; Bhatnagar, D.; Payne, G.A.; Nierman, W.C.; Bennett, J.W. Potential of *Aspergillus flavus* genomics for applications in biotechnology. *Trends Biotechnol.* **2009**, *27*, 151–157. [CrossRef] [PubMed]

56. Prakash, B.; Kedia, A.; Mishra, P.K.; Dubey, N.K. Plant essential oils as food preservatives to control moulds, mycotoxin contamination and oxidative deterioration of agri-food commodities—Potentials and challenges. *Food Control* **2015**, *47*, 381–391. [CrossRef]

57. Wang, M.; Jiang, N.; Xian, H.; Wei, D.; Shi, L.; Feng, X. A single-step solid phase extraction for the simultaneous determination of 8 mycotoxins in fruits by ultra-high performance liquid chromatography tandem mass spectrometry. *J. Chromatogr. A* **2016**, *1429*, 22–29. [CrossRef]

58. Rio, D.C.; Ares, M.; Hannon, G.J.; Nilsen, T.W. Purification of RNA using TRIzol (TRI Reagent). *Cold Spring Harb. Protoc.* **2010**, *6*, pdb–prot5439. [CrossRef]

59. Kim, D.; Langmead, B.; Salzberg, S.L. HISAT: A fast spliced aligner with low memory requirements. *Nat. Methods* **2015**, *12*, 357–360. [CrossRef]

60. Pertea, M.; Kim, D.; Pertea, G.M.; Leek, J.T.; Salzberg, S.L. Transcript-level expression analysis of RNA-seq experiments with HISAT, StringTie and Ballgown. *Nat. Protoc.* **2016**, *11*, 1650. [CrossRef]

61. Pertea, M.; Pertea, G.M.; Antonescu, C.M.; Chang, T.C.; Mendell, J.T.; Salzberg, S.L. StringTie enables improved reconstruction of a transcriptome from RNA-seq reads. *Nat. Biotechnol.* **2015**, *33*, 290. [CrossRef] [PubMed]

62. Buchfink, B.; Xie, C.; Huson, D.H. Fast and sensitive protein alignment using DIAMOND. *Nat. Methods* **2014**, *12*, 59. [CrossRef] [PubMed]

63. Conesa, A.; Götz, S.; García-Gómez, J.M.; Terol, J.; Talón, M.; Robles, M. Blast2GO: A universal tool for annotation, visualization and analysis in functional genomics research. *Bioinformatics* **2005**, *21*, 3674–3676. [CrossRef] [PubMed]

64. Finn, R.D.; Clements, J.; Eddy, S.R. HMMER web server: Interactive sequence similarity searching. *Nucleic Acids Res.* **2011**, *39*, W29–W37. [CrossRef] [PubMed]

65. Xie, C.; Mao, X.; Huang, J.; Ding, Y.; Wu, J.; Dong, S.; Kong, L.; Gao, G.; Li, C.Y.; Wei, L. KOBAS 2.0: A web server for annotation and identification of enriched pathways and diseases. *Nucleic Acids Res.* **2011**, *39*, W316–W322. [CrossRef] [PubMed]

66. Li, B.; Dewey, C.N. RSEM: Accurate transcript quantification from RNA-seq data with or without a reference genome. *BMC Bioinform.* **2011**, *12*, 323. [CrossRef] [PubMed]

67. Love, M.I.; Huber, W.; Anders, S. Moderated estimation of fold change and dispersion for RNA-seq data with DESeq2. *Genome Biol.* **2014**, *15*, 550. [CrossRef] [PubMed]

68. Klopfenstein, D.V.; Zhang, L.; Pedersen, B.S.; Ramírez, F.; Vesztrocy, A.W.; Naldi, A.; Mungall, C.J.; Yunes, J.M.; Botvinnik, O.; Weigel, M.; et al. GOATOOLS: A Python library for Gene Ontology analyses. *Sci. Rep.* **2018**, *8*, 10872. [CrossRef] [PubMed]

Article

Aspergillus flavus as a Model System to Test the Biological Activity of Botanicals: An Example on *Citrullus colocynthis* L. Schrad. Organic Extracts

Francesca Degola [1,*], Belsem Marzouk [2], Antonella Gori [3], Cecilia Brunetti [3,4], Lucia Dramis [1], Stefania Gelati [5], Annamaria Buschini [1,6] and Francesco M. Restivo [1]

[1] Department of Chemistry, Life Sciences and Environmental Sustainability, University of Parma, Parco Area delle Scienze 11/A, 43124 Parma, Italy; lucia.dramis@unipr.it (L.D.); annamaria.buschini@unipr.it (A.B.); restivo@unipr.it (F.M.R.)

[2] Laboratory of Chemical, Galenic and Pharmacological Development of Drugs, Faculty of Pharmacy of Monastir, University of Monastir, 5000 Monastir, Tunisia; belsemmarzouk@yahoo.fr

[3] Tree and Timber Institute (IVALSA), National Research Council of Italy (CNR), Via Madonna del Piano 10, 50019 Sesto Fiorentino, Firenze, Italy; antonella.gori@unifi.it (A.G.); cecilia.brunetti@unifi.it (C.B.)

[4] Department of Agriculture, Environment, Food and Forestry (DAGRI), University of Florence, Piazzale delle Cascine 18, 50144 Firenze, Italy

[5] Department of Packaging, Experimental Station for the Food Preserving Industry (SSICA), Viale Tanara 31/A, 43121 Parma, Italy; stefania.gelati@ssica.it

[6] Center for Molecular and Translational Oncology, Parco Area delle Scienze, 43124 Parma, Italy

* Correspondence: francesca.degola@unipr.it; Tel.: +39-0521-905603

Received: 15 April 2019; Accepted: 17 May 2019; Published: 22 May 2019

Abstract: *Citrullus colocynthis* L. Schrader is an annual plant belonging to the Cucurbitaceae family, widely distributed in the desert areas of the Mediterranean basin. Many pharmacological properties (anti-inflammatory, anti-diabetic, analgesic, anti-epileptic) are ascribed to different organs of this plant; extracts and derivatives of *C. colocynthis* are used in folk Berber medicine for the treatment of numerous diseases—such as rheumatism arthritis, hypertension bronchitis, mastitis, and even cancer. Clinical studies aimed at confirming the chemical and biological bases of pharmacological activity assigned to many plant/herb extracts used in folk medicine often rely on results obtained from laboratory preliminary tests. We investigated the biological activity of some *C. colocynthis* stem, leaf, and root extracts on the mycotoxigenic and phytopathogenic fungus *Aspergillus flavus*, testing a possible correlation between the inhibitory effect on aflatoxin biosynthesis, the phytochemical composition of extracts, and their in vitro antioxidant capacities.

Keywords: antimycotoxigenic activity; *Citrullus colocynthis*; *Aspergillus flavus*; model system; HPLC-MS/MS

---

1. Introduction

Oxidation is considered an underlying mechanism in the incidence of chronic diseases: Reactive oxygen species (ROS) such as superoxide anions, hydroxyl radicals, and hydrogen peroxide are cytotoxic, leading to tissue injuries. As in a "domino effect", oxidative stress resulting from the imbalance between the generation of reactive oxygen species and endogenous antioxidant systems induces inadvertent enzyme activation and consequent oxidative damage to cellular systems [1]. It is widely reported that cellular oxidative damage is responsible for numerous disorders, such as cardiovascular [2], Alzheimer's [3], and Parkinson's disease [4]—as well as ulcerative colitis [5], atherosclerosis [6], and cancer [7]. A key defense mechanism against radical mediated toxicity is represented by antioxidants, which protect cells from the damage caused by free radicals [8]. Consequently, during the last thirty years, several antioxidant-based formulations for the prevention and treatment of complex diseases have been developed [9–11]. Among these antioxidant formulations, a high number of plant-derived drugs, widely used for ethnopharmaceutical preparations, have been applied as "natural" principles included in modern medical applications [12,13]. Accordingly, interest in botanicals as a source of bioactive compounds has increased worldwide, and the finding of new biologically remarkable natural compounds affects not only the pharmaceutical field, but the nutraceutical and cosmeceutical fields too [14–17]. Clinical studies aimed at confirming the scientific bases (chemical and biological) of pharmacological activity of many plants used in folk medicine often rely on the results obtained from preliminary laboratory tests. For example, the antioxidant properties of plant extracts are typically evaluated through in vitro analyses, such as DPPH (2,2-diphenyl-1-picryl-hydrazyl-hydrate) and ABTS [2,2′-azinobis-(3-ethylbenzothiazoline-6-sulfonate)] assays [18], and to date, the number of in vitro studies far exceeds the number of in vivo studies, remaining the most cost effective and predominant type of research investigations performed. In fact, factors including costs, interspecific differences that preclude the adequate predictive value of the experiments, feasibility of testing procedures, and ethical concerns generally limit the utilization of animal models and human subjects for this kind of research [19]. The possibility of using high-throughput small-scale in vivo or ex vivo model systems to predict a possible antioxidant biological activity of plant extracts before their medical application is therefore desirable. Eukaryotic microorganisms, though few, have sometimes been selected for this purpose: For example, the yeast *Saccharomyces cerevisiae* has been employed to determine the antioxidant activity of different berry juices, which reportedly contain high amounts of phenolics [20].

Aspergillus flavus, a saprophytic plant pathogen, is the predominant species producing aflatoxins (AFs). Among multiple events that contribute to aflatoxin production, those involving ROS accumulation—such as during morphological and metabolic transitions—and the establishment of an oxidative intracellular environment seem to possess a key role in triggering AF biosynthesis. This correlation has been supported by the identification of at least one transcription factor, which responds to the cellular oxidative stress by activating a series of enzymes responsible for the scavenging of cytoplasmic ROS excess [21–23]. Conversely, many compounds with antioxidant properties (such as ascorbate, eugenol, ethylene, methyl jasmonate, and α-lipoic acid) showed to exert an inhibition/containment effect on AF biosynthesis [24–27], mainly through the stimulation of catalase, superoxide dismutase, and glutathione peroxidase activity [28]. An inhibitory effect on AF biosynthetic pathway and *A. flavus* growth has been recently reported for a wide number of botanicals and essential oils [29–31]. Since the response of aflatoxin metabolism to redox balance alterations is well known, as they are considered a sort of "defense molecule" synthesized to cope with an excess of ROS in the late phase of growth [22,23], we proposed this airborne microorganism as a model system to screen the antioxidant potential of plant extracts.

Here we analyzed the effect of organic extracts of *Citrullus colocynthis* L. Schrader, an annual plant belonging to the Cucurbitaceae family which grows in arid and semi-arid regions, on AF biosynthesis and *A. flavus* growth. Native to tropical Asia and Africa, *C. colocynthis* is is now widely distributed in the desert areas of the Mediterranean basin (in Italy the only known population is located in the Aeolian island of Vulcano). Many pharmacological properties (anti-inflammatory, anti-diabetic, analgesic, anti-epileptic) are ascribed to different organs of this plant [32–35]: Extracts and derivatives of *C. colocynthis* are used in folk Berber medicine for the treatment of numerous diseases; the root is used for arthritic pain, breast inflammation, ophthalmia, and uterine pain; and the leaves are used for treatment of cough, many tumors, and as a cholagogue [36]. Recently, antifungal and antibacterial activities of organic extracts from leaves and seeds were reported [37–39]. However, despite various studies on the medical use of *C. colocynthis* derivatives, information about its antimycotoxigenic potential are still scarce. The aim of this work is to evaluate the antiaflatoxigenic effect of organic extracts of *C. colocynthis* stem, leaf, and root through the use of *A. flavus* as a model system, comparing their effect on the basis of phytochemical composition.

2. Results and Discussion

2.1. Phytochemical Characterization of C. colocynthis Extracts

Several studies reported a high in vitro antioxidant potential of organic *C. colocynthis* extracts obtained from various tissues, due to their polyphenolic composition [40,41]. The antioxidant capacity of root, stem, and leaf extracts and phenolic content were then measured according to the DPPH and Folin-Ciocalteu's methods, respectively. As a general consideration, it should be noted that stem and leaf extracts showed a wider range of antioxidant activity, depending on the extraction solvent, than root extracts (Figure 1). The highest antioxidant capacity was determined in the methanol (MET) leaf extract and ethyl acetate (EA) root extracts, followed by chloroform (CHL) and methanol root extracts. On the contrary, MET stem extracts showed the lowest activity (Figure 1).

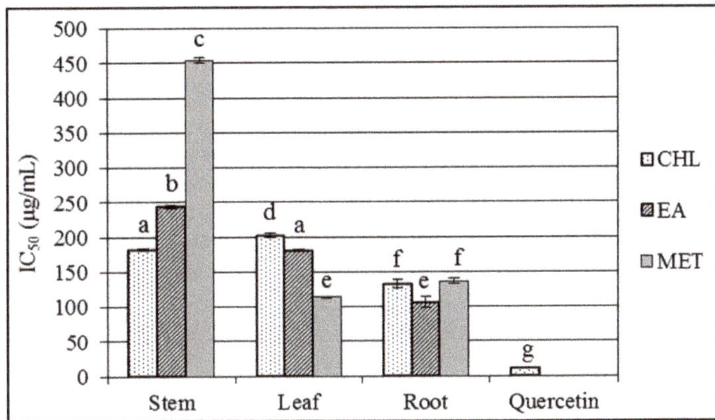

Figure 1. DPPH scavenging activity of *Citrullus colocynthis* root, leaf, and stem extracts. Data are means of three replicates ± S.D. Same letters indicate absence of statistically significant differences ($p < 0.05$).

A detailed phytochemical characterization of the leaves, stems, and roots of *C. colocynthis* revealed the presence of different classes of metabolites—such as coumarins, hydroxycinnamic acid derivatives, flavan-3-ols glycosides, flavone glycosides, and tetracyclic triterpenes (Table 1)—as according to previous investigations [42–44]. These compounds, which have been indicated as responsible for antifungal activity against *Aspergillus* strains [38], were abundant in the analyzed extracts (Table 1). Esculetin (1), p-coumaric acid derivatives (2, 5), orientin (3), vitexin (6), apigenin derivatives (4, 7, 8, 9, 12, 17), and epicatechingallate (18) were identified according to their MS fragmentation pattern and absorption spectra (Table 1). The remaining identified peaks corresponded to different flavone derivatives (10, 11), and cucurbitacin derivatives (13, 14, 15, 16) were elucidated by their molecular weight obtained by MS analysis, their UV spectra, and by comparing experimental data with respective literature data (Table 1) [45–47]. In particular, cucurbitacin derivatives and colocynthoside B did not furnish any MS fragment as previously reported by Chawech et al. [44]. In order to exclude that these compounds were artefacts due to solvent extraction, they were also compared with MS and UV spectra of authentic standards of cucurbitacin E and I. Compound **19** (Table 1) was tentatively identified as colocynthoside B on the basis of its molecular weight, its UV spectrum, and the comparison with literature data [48]. Marked differences were observed among the tissues. However, in all three extracts, the leaf contained the highest variety of phenolics, as well as the higher content of secondary metabolite, except for compound **12** (Table 1), which was more abundant in the ethyl acetate extract of the stem. Overall, our results showed that ethyl acetate was the most efficient solvent to extract phenolic constituents in all tissues, apart from orientin, coumaric flavone derivative, two apigenin hexosides (compounds **7** and **17**), epicatechin gallate, vitexin, and compound **5**—which were more concentrated in methanol extracts. On the contrary, chloroform was efficient to extract cucurbitacin I derivatives (compounds **13**, **14**, **16**) and colocynthoside B. The most abundant cucurbitacin derivative (compound **13**) resulted in extraction to the same extent as in ethylacetate and chloroform. The presence of cucurbitacins is relevant to the bitterness and toxicity of the plant, but it also has some biological effects—such as anti-inflammatory, purgative, and anti-cancer activities [49].

2.2. Antifungal and Anti-Aflatoxigenic Activity

The prediction of the biological activity of natural extracts may often be difficult due to the variation in their chemical constituents, that in turn depend on the growth stages of the plants, and/or their geographic origin [50]. On the other hand, screening plant crude extracts can simplify the discovery of new and promising bioactives, and allows a further, more specific identification of the chemical compounds responsible for the observed activity. We performed a preliminary assay to evaluate the effect of the different *C. colocynthis* extracts on *A. flavus* growth, intended as daily radial increase of fungal colonies diameter. Concentration of 500 µg/mL was tested for each extract. As reported in Table 2, none of them resulted in a significant reduction of radial mycelium growth, suggesting that the composition of extracts did not possess appreciable antimicrobial activity against the fungus.

These preliminary results led to the exclusion of any antifungal or fungistatic potential of extracts on the mycelium long-term growth. However, various studies have reported that several compounds, both synthetic and natural, are effective in lowering AF production without apparently interfering with the fungal development [51,52]. At present, these compounds are a promising tool for uncovering the regulatory mechanisms triggering the mycotoxigenic metabolism, one of the main targets for mycotoxin diffusion/contamination control strategies [53,54].

To assess and compare the efficacy of different organic extracts of *C. colocynthis* tissues on total AF biosynthesis, conidia of *A. flavus* were inoculated in clarified coconut medium (CCM) and the fluorescence-based microplate procedure was used [27]. Leaf, stem, and root organic extracts were tested at increasing concentrations (data not shown); the lower and the higher concentrations (100 and 500 µg/mL, respectively) are reported in Figure 2. Chloroform (CHL), ethyl acetate (EA), and methanol (MET) extracts were administrated to aflatoxigenic *A. flavus* cultures. After six days of incubation, aflatoxin accumulation showed a dose-dependent alteration in response to extract exposure: The lowest dose (100 µg/mL) was less effective in limiting the amount of toxins in the culture, while the effect increased when extracts were added at the concentration of 500 µg/mL. Among tissues, leaf and root extracts had the highest levels of aflatoxin inhibition (Figure 2B,C), exceeding 80% inhibition in both CHL extracts. In addition, root extracts were able to lower the aflatoxin concentration by up to 12% in the most effective (CHL), and around 45% in the case of the least effective (MET; Figure 2B).

Table 1. UPLC-DAD-MS/MS characterization and quantification (μmol g^{-1} D.W.) of main secondary metabolites present in the different extracts of *C. colocynthis*. Data are means $\pm$ S.D. ($n = 3$). * nd = not detectable.

n°	Name	MW	[M − H]⁻, *m/z*	MS/MS, *m/z*	λmax, nm	Ethyl Acetate			Methanol			Chloroform		
						Leaf	Stem	Root	Leaf	Stem	Root	Leaf	Stem	Root
1	Esculetin	178	177	133, 105, 89	330	40.6 ± 4.91	10.5 ± 1.08	17.7 ± 2.03	n.d.	n.d.	n.d.	n.d.	n.d.	n.d.
2	p-Coumaric acid	164	163	147, 119	225, 310	20.7 ± 1.24	6.55 ± 0.45	1.67 ± 0.09	1.59 ± 0.12	1.75 ± 0.17	n.d.	n.d.	n.d.	n.d.
3	Orientin	448	447	327, 357, 285	350, 269	n.d.	n.d.	0.28 ± 0.01	140.3 ± 20.87	7.30 ± 0.97	1.20 ± 0.18	n.d.	n.d.	n.d.
4	Apigenin-hexoside	432	431	311, 269, 211, 159	267, 337	1.59 ± 0.23	0.41 ± 0.05	n.d.	n.d.	n.d.	n.d.	n.d.	n.d.	n.d.
5	trans p-Coumaric acid 4-*O*-malate	280	279	147, 119	225, 310	0.79 ± 0.09	1.89 ± 0.31	n.d.	2.837 ± 0.21	5.44 ± 0.61	n.d.	n.d.	n.d.	n.d.
6	Vitexin	432	431	311, 341, 269	267, 337	3.86 ± 0.04	n.d.	n.d.	7.69 ± 0.68	4.66 ± 0.59	n.d.	n.d.	n.d.	n.d.
7	Apigenin -D-glucopyranosyl-8-apiofuranoside	564	563	311, 269	267, 337	59.6 ± 0.73	13.10 ± 1.44	n.d.	529.1 ± 30.17	108.5 ± 11.39	n.d.	n.d.	n.d.	n.d.
8	Apigenin derivative (isomer 1)	548	547	311	267, 337	13.5 ± 1.70	3.16 ± 0.29	0.05 ± 0.00	4.6 ± 0.42	0.31 ± 0.00	0.7	n.d.	n.d.	n.d.
9	Apigenin derivative (isomer 2)	548	547	311	267, 337	47.7 ± 3.99	14.3 ± 1.74	0.28 ± 0.01	13.6 ± 0.15	8.09 ± 0.06	1.77	n.d.	n.d.	n.d.
10	caffeoyl malic Flavone	578	577	179	340, 270	3.08 ± 0.29	1.19 ± 0.02	n.d.	n.d.	n.d.	n.d.	n.d.	n.d.	n.d.
11	coumaric Flavone derivative	584	583	285, 147	348, 310	4.88 ± 0.51	5.06 ± 0.61	0.08 ± 0.00	55.1 ± 6.48	6.68 ± 0.52	1.34 ± 0.14	n.d.	n.d.	n.d.
12	Apigenin derivative	752	751	311	267, 337	n.d.	63.6 ± 5.41	24.9 ± 1.84	12.8 ± 2.54	n.d.	n.d.	n.d.	n.d.	n.d.
13	Cucurbitacin E	556	555	n.d.	229	54.2 ± 6.0	57.9 ± 6.9	13. 6 ± 0.98	n.d.	6.84 ± 0.54	2.48 ± 0.54	44.5 ± 5.41	55.9 ± 7.82	14.5 ± 2.54
14	Cucurbitacin I	514	513	n.d.	229	99.8 ± 8.21	5.81 ± 0.42	27.6 ± 3.41	n.d.	n.d.	n.d.	212.6 ± 32.77	220.2 ± 19.9	41.3 ± 6.77
15	acetyl Cucurbitacin E	760	759	n.d.	229	9.31 ± 0.57	n.d.	3.59 ± 0.08	n.d.	n.d.	n.d.	n.d.	n.d.	n.d.
16	coumaroyl acetyl Cucurbitacin I	864	863	n.d.	229	7.13 ± 0.44	n.d.	0.99 ± 0.00	n.d.	n.d.	n.d.	36.6 ± 4.09	n.d.	n.d.
17	Apigenin-dihexoside	594	593	311	267, 337	n.d.	n.d.	n.d.	15.4 ± 2.01	18.1 ± 2.55	n.d.	n.d.	n.d.	n.d.
18	Epicatechin gallate	442	441	289	280	n.d.	n.d.	n.d.	16.8 ± 2.11	10.8 ± 0.98	n.d.	n.d.	n.d.	n.d.
19	Colocynthoside B	806	805	n.d.	230	n.d.	n.d.	n.d.	n.d.	n.d.	n.d.	25.6 ± 3.29	3.89 ± 0.45	6.69 ± 0.71

Table 2. Effect of 500 µg/mL *C. colocynthis* extracts on *A. flavus* radial growth. Radial increment is expressed as the mean of daily radial increase of colonies radius (cm/d) ± S.D. Same letters indicate absence of statistically significant differences ($p < 0.05$).

	CHL	EA	MET	CNT
Root	0.45 ± 0.13 [a]	0.43 ± 0.06 [a]	0.43 ± 0.09 [a]	0.47 ± 0.08 [a]
Stem	0.45 ± 0.11 [a]	0.39 ± 0.15 [a]	0.44 ± 0.12 [a]	0.47 ± 0.08 [a]
Leaf	0.44 ±0.13 [a]	0.46 ± 0.08 [a]	0.43 ± 0.10 [a]	0.47 ± 0.08 [a]

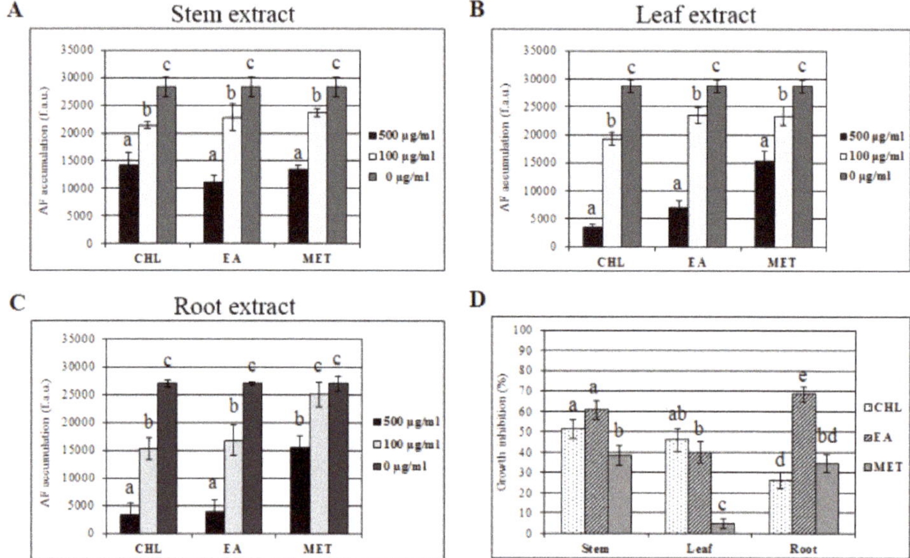

Figure 2. Activity on toxin accumulation and mycelium growth. Aflatoxin accumulation (reported as fluorescence arbitrary units) in *A. flavus* six-days after CCM cultures treated with stem (**A**), leaf (**B**), and root (**C**) extracts. (**D**) Early mycelium growth inhibition of *A. flavus* conidia treated with 500 µg/mL extracts, measured 48 h after inoculum by optical density increasing, and expressed as percentage in respect to control. Error bars refer to mean values of four replicates ± S.D.

A similar correlation between solvent and aflatoxin inhibition rate was observed for the highest concentration of EA leaf and root extracts in eculetin, p-coumaric acid, apigenin, and cucurbitacin derivatives (Table 1). Interestingly, the high activity of the CHL extract could be related to cucurbitacin derivatives and colocynthoside B. In particular, colocynthoside B was detected only in the chloroform extracts and may be responsible of the observed aflatoxin inhibition (Figure 2). Prevention of AF accumulation has, for a long time now, been associated with the action of molecules and/or conditions that interfere with fungal growth [24,29,30]; however, during the last decades, several other substances have been shown to be effective in completely blocking the biosynthesis of mycotoxins without affecting mycelium development [52,55]. Thus, the correlation between aflatoxin metabolism and *A. flavus* growth should be considered under different perspectives of fungal development: Colonies may have the same radius, but vary significantly in hyphal density and, therefore, biomass. In fact, hyphae branching, which is necessary for an efficient colonization and utilization of the substrate, responds to nutrient gradients, growing away from areas staled by metabolic by-products of existing hyphae. However, colony radial growth is not influenced by the concentration of nutrients, since existing hyphal tips at the colony margin, which determine the colony diameter, have priority over all other hyphal tips (i.e., the branches). For this reason, the evaluation of colony radial growth as a unique parameter for the assessment of any antifungal effect could be misleading about possible fungistatic activities of tested compound/mixture, or might disguise an early stadium effect. As reported in Figure 2D, when *C. colocynthis* extracts were administrated to *A. flavus* conidia in YES liquid cultures at the higher concentration (500 µg/mL), early mycelium development was delayed by the majority of extracts. The exerted effect depended either on the tissue or solvent: For example, while CHL and EA extracts from the stem and leaf did not significantly differ in their inhibitory effect (50 vs. 60% and 45 vs. 50%, respectively), EA root extract proved to be more highly effective against the initial development of mycelium than CHL extract from the same tissue (70 vs. 25%).

2.3. Aflatoxin-Modulating Activity

Time-dependent aflatoxin production was analyzed by time-course experiments, where the kinetic of toxin accumulation was "real time", determined by starting from 65 to 146 h after inoculum in CCM medium. As previously reported [27] the AF concentration in the control cultures progressively increased for up to 85–90 h, maintaining, from here on, a 'plateau' value. AF production in the *A. flavus* cultures treated with a 500 µg/mL extract concentration showed a similar time course, but the maximum quantity of toxins produced varied consistently, with the extract-dependent inhibition rate observed in the end-point cultures. Stem extracts did not significantly differ from each other in terms of global inhibition level, blocking AF accumulation to 50% of the control value (Figure 3A). On the contrary, leaf and root extracts resulted in a variable range of toxin containment: The AF accumulation course, in cultures treated with leaf extracts, split from control at 72 h after inoculum, reaching a peak at 89 h (Figure 3B). A similar pattern was observed in root MET treated cultures, whereas root CHL and root EA avoided AF accumulation already before 65 h (Figure 3C).

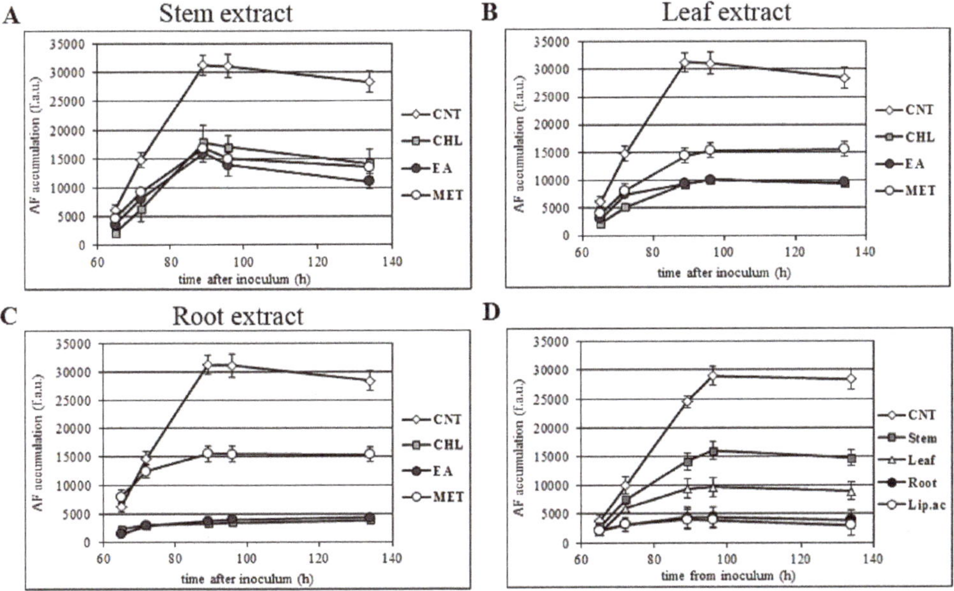

Figure 3. Aflatoxin time-course accumulation. Effect of stem (**A**), leaf (**B**), and root (**C**) extracts on AF time-course accumulation in *A. flavus*. (**D**) Comparison between 500 µg/mL stem, leaf, and root CHL extract; α-lipoic acid 1 mM is used as a reference. Error bars refer to mean values of four replicates ± SD.

Due to the variety of synergistic phenomena occurring in the cell, the antioxidant activity (and resulting biological effect) may rarely be calculated on the basis of chemical in vitro assays, mainly in the case of botanicals and phytocomplexes as those reported here. Therefore, the effect of different extracts on the redox balance could only in part be predicted. Additionally, the use of high concentrations of single constituents (when, actually, a plant extract is a complex mixture) may result in the in vitro system being exposed to an overstated and unrealistic concentration. However, a comparison with a standard molecule, owning well-documented antioxidant properties consistent with the biological effect in the object of this study, should be done. Lipoic acid is a well-known ROS scavenger, and its efficacy in preventing aflatoxin production in *A. flavus* was already reported [27]. At the highest dose considered here (500 µg/mL), CHL root extracts resulted in AF accumulation containment, which was comparable to that observed for 1 mM α-lipoic acid. Stem and leaf CHL extracts were less efficient inhibitors of AF biosynthesis, as compared to α-lipoic acid and root CHL extracts. It thus appears that CHL root extracts are a promising source of anti-aflatoxigenic molecules. On the other hand, reinforcing evidence shows that *A. flavus* aflatoxin-producing strains may be used as an in vivo model to test the antioxidant activity of new mixture/compounds.

2.4. Time Course of Extract Administration on Aflatoxin Accumulation

According to various authors, many plant extracts showing an inhibitory effect against aflatoxin accumulation at the early stage seemed to become almost ineffective after protracted incubation [56,57], suggesting that their biological activity might depend not only on phytochemical composition, but also on the chemical structure and related properties of single components. Additionally, evidence has been provided showing that the delivery time of an anti-oxidant compound during fungal growth may affect its inhibitory efficacy on aflatoxin biosynthesis and/or accumulation [40]. In Figure 4, the time course of *C. colocynthis* root extracts (CHL and EA) administration is reported. As a general observation, early administration (time 0; at the germination stage) of the relevant extract in the medium resulted in the highest inhibition of AF accumulation, whereas delaying the time of administration (from 65 h onwards) did not block mycotoxin biosynthesis but, in the best case (65 h), retarded aflatoxin accumulation.

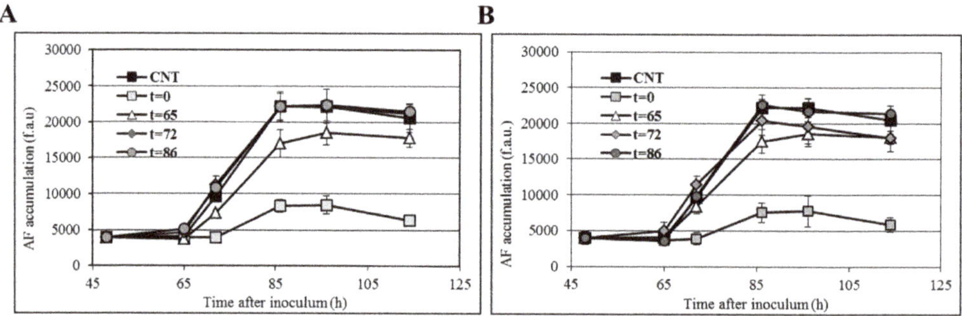

Figure 4. Effect of 500 µg/mL CHL (**A**) and EA (**B**) root extract over time on aflatoxin production. Extracts were added to conidia of *A. flavus* inoculated in CCM after 65, 72, and 86 h of incubation. Error bars refer to mean values of four replicates ± SD.

As previously reported for lipoic acid [27], the efficacy of CHL and EA root extracts in preventing AF accumulation is limited to a short time interval (0–65 h) that precedes the burst of AF biosynthesis. It would be worth analyzing the metabolic and regulatory networks operating during this time window to uncover possible molecular targets for designing new and specific anti-aflatoxigenic compounds.

2.5. Conidia Production and Conidiophores Morphology

In Aspergilla, vegetative reproduction and subsequent colonization of the surrounding environment rely on the differentiation of specialized structures (conidiophores) bearing vegetative spores (conidia), whose formation process characterizes the late phase of mycelium growth. We tested the effect of *C. colocynthis* CHL and MET stem, leaf, and root extracts on the production of conidia; as reported in Figure 5, a tissue and solvent (CHL and MET) dependent efficacy of the various extracts on lowering the number of conidia accumulated by treated mycelia was observed.

The overall pattern of tissue/solvent efficacy on conidia production was quite different with that observed for mycelium growth (Table 2) or hyphae elongation, and for AF accumulation (Figure 2). Interestingly, the scanning emission microscopy analysis (SEM) conducted to evaluate the conidiophore organization showed that no significant alteration of either the morphology or the general aspect of these reproductive structures occurred (an example is reported in Figure 5B), providing evidence that the relevant treatment affected the number of conidia or conidiophores.

Figure 5. Effect of 500 µg/mL root, stem, leaf (R, S, L) MET and CHL extracts on conidia production (**A**) and conidiophores morphology (**B**) (left: Control; right: R-CHL treated cultures) in *A. flavus* 96-wells CCM cultures. Values are reported as inhibition percentages in respect to control; Error bars refer to mean values of four replicates ± S.D. Different letters over the bars indicate the differences that were statistically significant ($p < 0.05$).

3. Conclusions

Plants with significant pharmacological properties have often been found to be rich in polyphenols and other secondary metabolites that have been proven to possess high antioxidant potentials due to their activity as reducing agents, metal chelators, and free radical quenchers. In this sense, every bioactive able to interfere with the oxidative status of fungal cells, on which the mycotoxin metabolism relies, should be validated as a biocontrol agent in organic strategies aimed at reducing aflatoxin contamination in food and feed commodities. On the other hand, a search for new antifungal and antimycotoxigenic substances is of increasing interest, with the perspective of improving antifungal resistance and understanding the underlying mechanisms of these new drugs with a wide range of applications—from medical mycology to agricultural and food safety.

4. Materials and Methods

4.1. Plant Materials

Citrullus colocynthis L. Schrader plants were collected near Medenine (Tunisia), in the municipality of Sidi Makhlouf. The identification was performed according to the flora of Tunisia [58] and a voucher specimen (C.C-01.01) deposited in the biological laboratory of the Faculty of Pharmacy of Monastir.

4.2. Extraction Protocol

Fresh tissues (roots, stems, and leaves) were dried and powdered using a tissue blender. Different solvents, in ascending polarity (petroleum ether, chloroform, ethyl acetate, and methanol) were used for Soxhlet extraction to fractionate the soluble compounds from the plant material. The extraction was performed with dried powder (100 g) placed inside a thimble made by thick filter paper, loaded into the main chamber of the Soxhlet extractor, which consisted of an extracting tube, a glass balloon, and a condenser. The total extracting time was 6 h for each solvent, continuously refluxing over the sample at a temperature not exceeding the boiling point. The resulting extracts were evaporated at reduced pressure to obtain the crude extracts. The organic solvents used were 99% pure. Extracts were all ethanol resuspended for further analysis. All the chemicals were obtained from Sigma (St. Louis, MO, USA).

4.3. Determination of the Total Phenolic Contents

Phenolic compound concentration in the different extracts was determined by using the Folin–Ciocalteu's phenol reagent, according to Singleton and Rossi [59], with some modifications. Briefly, 100 µL of the extract solution was mixed with 100 µL of Folin–Ciocalteu's phenol reagent. After 3 min, 100 µL of saturated sodium carbonate solution was added to the mixture and adjusted to 1 mL with distilled water. The reaction was kept in the dark for 90 min, after which the absorbance was recorded at 720 nm. Gallic acid was used to design the standard curve. The contents of total phenolic are expressed as mg of gallic acid equivalents (GAE)/g of extract. Data were reported as means of three replicates ± S.D.

4.4. Determination of DPPH Radical Scavenging Activity

The ability to scavenge the DPPH-free radical was monitored according to a method first introduced by Blois (1958) and developed by Brand-Williams et al. [60]. Various concentrations of sample extracts (0.5 mL) were mixed with 0.5 mL of methanolic solution containing DPPH radicals (6×10^{-5} M). The mixture was shaken vigorously and left to stand in the dark until stable absorption values were obtained. The reduction of the DPPH radical was measured by continuously monitoring the decrease of absorption at 517 nm. The DPPH scavenging effect was calculated as a percentage of DPPH discoloration using the following equation: % scavenging effect = [(ADPPH × AS)/ADPPH] × 100, where AS is the absorbance of the solution when the sample extract has been added at a particular level, and ADPPH is the absorbance of the DPPH solution. Three experiments were performed in triplicate. The antiradical activity was expressed in terms of the amount of antioxidant necessary to decrease the initial DPPH absorbance by 50% (IC50). The IC50 value for each extract was determined graphically by plotting the percentage of DPPH scavenging as a function of extract concentration.

4.5. UPLC-DAD-ESI-MS/MS Analysis

For the chemical characterization, 60 mg of each extract were re-dissolved in methanol-distilled water (1:1 *v/v*) and filtered with a PTFE membrane. 5 µL were injected in a LC–DAD-MS/MS system, consisting of a Shimadzu Nexera UPLC system (Kyoto, Japan) coupled with a diode array detector (DAD), and a Shimadzu LCMS-8030 quadrupole mass spectrometer (Kyoto, Japan) equipped with an electrospray ionization source (ESI). Analytical separation was performed on a reversed-phase Waters

Nova-Pak C18 column (4.9 × 250 mm, 4 µm) (Water Milford, MA, USA), operating at 30 °C. The mobile phase consisted of 1% aqueous formic acid (solvent A) and 1% formic acid in acetonitrile (solvent B). The elution gradient consisted of 3% B isocratic for 5 min, from 5 to 100% B linear for 30 min, 100% B isocratic for 7 min. The flow rate was 0.5 mL/min. The mass spectrometer operated in Negative Ion Scan and in Product Ion Scan mode, acquiring over a mass-range from m/z 50 to 1100 and using Argon as Collision Induced Dissociation (CID) gas at a pressure of 230 kPa. The interface voltage was set to −3.5 kV; desolvation line (DL) temperature was 250 °C and the heat block temperature was 400 °C.

Identification of the major secondary metabolites in the different extracts was carried out using their retention times, and both UV-VIS, MS and MS/MS spectra. Quantification of single compounds was performed by UPLC-DAD in triplicates through an external standard method, using stock solutions of the following compounds: Esculetin, p-coumaric acid, orientin, catechin, vitexin, (all from Sigma-Aldrich, Milan, Italy) and cucurbitacin I, cucurbitacin E, and apigenin-7-O-glucoside from Extrasynthese (Lyon, France). All solvents used for the analyses were purchased from Sigma-Aldrich (Milan, Italy).

4.6. Fungal Strains, Media and Culture Condition

Aspergillus flavus toxigenic strain Fri2 and atoxigenic strain TOφ used were previously isolated from corn fields of the Po Valley [61]. Conidia suspensions were obtained from 10-day YES-agar [2% (*w/v*) yeast extract (Difco, Detroit, MI, USA), 5% (*w/v*) sucrose (Sigma, St Louis, MO, USA), 2% (*w/v*) agar (Difco)] cultures incubated at 28 °C; conidia concentration (quantified by OD_{600}) and viability (>90%) were determined according to Degola et al. [42]. Coconut milk-derived medium (CCM) used for microplate assays was obtained as described in Degola et al. [27]: Briefly, 400 mL of commercial coconut cream was diluted to the final volume of 1.2 L with bidistilled water, sterilized by autoclaving (10 min, 120 °C), cooled at 4 °C overnight, and clarified by centrifugation (15 min at 3200× *g*). The residual floating material and the pellet were discarded, and the intermediate phase was then recovered and used as culture medium in the aflatoxin inhibition assays.

4.7. Aflatoxin Production Assay

The extracts' effects on aflatoxin biosynthesis were assessed by the microplate fluorescence-based procedure described in Degola et al. [27]. Standard flat-bottom 96-well microplates (Sarstedt, Newton, NC, USA) were used. Suspensions of conidia were diluted to the appropriate concentrations and brought to the final concentration of 5×10^2 conidia/well; cultures were set in a final volume of 200 µL/well of CCM medium. *C. colocynthis* organic extracts were ethanol resuspended and added to the culture medium. The plates were incubated in the dark under stationary conditions for up to 6 days at 25 °C; visual inspection of mycelium development and conidiation served as an indicator of the culture growth. Total aflatoxin accumulation was monitored by fluorescence emission determination; readings were performed directly from the wells bottom of the culture plate with a microplate reader (TECAN SpectraFluor Plus, Männedorf, Switzerland) using the following parameters: λ_{ex} = 360 nm; λ_{em} = 465 nm; manual gain = 83; lag time = 0 µs; number of flashes = 3; integration time = 200 µs). Inocula were performed in quadruplicate.

4.8. Aspergillus Flavus Growth

A. flavus radial growth was performed in YES-agar added with *C. colocynthis* extracts at 500 µg/mL final concentration: Three equidistant single spots (5 µL of a 10^7 conidia/mL suspension each) of aflatoxigenic strain Fri2 were inoculated in Petri dishes (9 cm Ø), plates were incubated for 4 days at 25 °C, and the mycelium growth was evaluated daily by measuring colonies reverse along two orthogonal diameters. YES-agar plates supplemented with 0.5% EtOH (*v/v*) were used as control. Early mycelium development was assessed, recording changes in optical density of liquid cultures over time: In a 96 wells microtiter plate (Sarstedt, Newton, NC, USA), 1×10^4 conidia were inoculated in a final volume of 200 µL of YES 5% liquid medium, added with 500 µg/mL organic extracts, and incubated at

28 °C. The optical density at 620 nm (OD_{620}) was recorded for each well, using a microplate reader (RosysAnthos ht3; AnthosLabtec Instruments GmbH, Salzburg) without shaking. Experiments were performed in quadruplicate.

4.9. Conidiation Rate Assessment and Reproductive Structures Analysis

Conidia production was estimated for CHL and MET leaf, stem, and root extract treated cultures (500 µg/mL). From the eight replicates of each condition, four mycelia were collected from the CCM microplates used in the AF accumulation assay, and individually washed three times in a 0.1% (*v/v*) Tween20® aqueous solution by vortexing 1 min. The spore suspensions were then washed three times with a 80% (*v/v*) ethanol solution and conidia concentration was then determined with a Burker chamber. The remaining four CHL and MET leaf, stem, and root extract treated cultures from AF accumulation plates were observed using a Scanning Electron Microscope (SEM) JEOL IT 300, in high vacuum mode. Samples were fixed for 4 h in 3% (*v/v*) glutaraldehyde, acetone dehydrated (from 30% to 100% water/acetone solutions), critical point dried, and coated with a thin layer of gold by means of a Sputter Coater "Agar". Observations were conducted at an acceleration voltage of 10.0 kV and at a 450× magnification.

4.10. Statistical Analysis

The data were analyzed using the statistical and graphical function of PASW Statistics (SPSS Inc., Chicago, IL, USA). Differences were assessed using analysis of variance (ANOVA), followed by Dunnet-t post hoc test.

Author Contributions: Conceptualization, F.D., F.M.R., A.B., and B.M.; Investigation and Visualization, F.D.; Chemical Analyses, B.M., C.B., and A.G.; Electron Scanning Microscopy Analysis, L.D., F.D., and S.G.; Resources, B.M., F.M.R., C.B., and S.G.; Writing—Original Draft Preparation, F.D. and B.M.; Writing—Review & Editing, F.M.R., A.B., C.B., and A.G.

Funding: This research received no external funding.

Acknowledgments: We are indebted to the Laboratoire de biologie végétale, Unité de Pharmaco-économie et développement des médicaments, Faculté de Pharmacie, Monastir (Tunisia) for providing the *C. colocynthis* organic extracts. We are also indebted to Massimo Cigarini for the technical help. We would like to thank Prof. Justice Johannson for the English language editing of proof.

Conflicts of Interest: The authors declare no conflict of interest.

References

1. Wiseman, H.; Halliwell, B. Damage to DNA by Reactive Oxygen and Nitrogen Species: Role in Inflammatory Disease and Progression to Cancer. *Biochem. J.* **1996**, *313*, 17–29. [CrossRef] [PubMed]
2. Singh, U.; Jialal, I. Oxidative stress and atherosclerosis. *Pathophysiology* **2006**, *13*, 129–142. [CrossRef]
3. Smith, M.A.; Rottkamp, C.A.; Nunomura, A.; Raina, A.K.; Perry, G. Oxidative stress in Alzheimer's disease. *Biochim. Biophys. Acta* **2000**, *1502*, 139–144. [CrossRef]
4. Onyou, H. Role of Oxidative Stress in Parkinson's Disease. *Exp. Neurobiol.* **2013**, *22*, 11–17.
5. Ramakrishna, B.S.; Varghese, R.; Jayakumar, S.; Mathan, M.; Balasubramanian, K.A. Circulating antioxidants in ulcerative colitis and their relationship to disease severity and activity. *J. Gastroenterol. Hepatol.* **1997**, *12*, 490–494. [CrossRef] [PubMed]
6. Kattoor, A.J.; Pothineni, N.V.K.; Palagiri, D.; Mehta, J.L. Oxidative Stress in Atherosclerosis. *Curr. Atheroscler. Rep.* **2017**, *19*, 42. [CrossRef]
7. Kinnula, V.L.; Crapo, J.D. Superoxide dismutases in malignant cells and human tumors. *Free Radic. Biol. Med.* **2004**, *36*, 718–744. [CrossRef]
8. Lien Ai, P.-H.; Hua, H.; Chuong, P.-H. Free Radicals, Antioxidants in Disease and Health. *Int. J. Biomed. Sci.* **2008**, *4*, 89–96.
9. Serafini, M.; Peluso, I. Functional Foods for Health: The Interrelated antioxidant and anti-inflammatory role of rruits, vegetables, herbs, spices and cocoa in humans. *Curr. Pharm. Des.* **2016**, *22*, 6701–6715. [CrossRef]

10. Davatgaran-Taghipour, Y.; Masoomzadeh, S.; Farzaei, M.H.; Bahramsoltani, R.; Karimi-Soureh, Z.; Rahimi, R.; Abdollahi, M. Polyphenol nanoformulations for cancer therapy: Experimental evidence and clinical perspective. *Int. J. Nanomed.* **2017**, *12*, 2689–2702. [CrossRef]

11. Ahmad, N.; Mukhtar, H. Antioxidants meet molecular targets for cancer prevention and therapeutics. *Antioxid Redox Signal.* **2013**, *19*, 85–88. [CrossRef]

12. Zhang, Y.J.; Gan, R.Y.; Li, S.; Zhou, Y.; Li, A.N.; Xu, D.P.; Li, H.B. Antioxidant phytochemicals for the prevention and treatment of chronic diseases. *Molecules* **2015**, *20*, 21138–21156. [CrossRef]

13. Szymanska, R.; Pospisil, P.; Kruk, J. Plant-Derived Antioxidants in Disease Prevention. *Oxidative Med. Cell. Longev.* **2016**, *2016*, 192–208. [CrossRef]

14. Naveen, J.; Baskaran, V. Antidiabetic plant-derived nutraceuticals: A critical review. *Eur. J. Nutr.* **2018**, *57*, 1275–1299. [CrossRef]

15. Tessema, E.N.; Gebre-Mariam, T.; Neubert, R.H.H.; Wohlrab, J. Potential Applications of Phyto-Derived Ceramides in Improving Epidermal Barrier Function. *Ski. Pharm. Physiol.* **2017**, *30*, 115–138. [CrossRef]

16. Pandey, K.B.; Rizvi, S.I. Plant polyphenols as dietary antioxidants in human health and disease. *Oxidative Med. Cell. Longev.* **2009**, *2*, 270–278. [CrossRef]

17. Reuter, J.; Merfort, I.; Schempp, C.M. Botanicals in dermatology: An evidence-based review. *Am. J. Clin. Dermatol.* **2010**, *11*, 247–267. [CrossRef]

18. Fidrianny, I.; Rahmiyani, I.; Wirasutisna, K.R. Antioxidant capacities from various leaves extracts of four varieties mangoes using DPPH, ABTS assays and correlation with total phenolic, flavonoid, carotenoid. *Int. J. Pharm. Pharm. Sci.* **2013**, *5*, 189–194.

19. Markowitz, J.S.; Zhu, H.J. Limitations of in vitro assessments of the drug interaction potential of botanical supplements. *Planta Med.* **2012**, *78*, 1421–1427. [CrossRef]

20. Slatnar, A.; Jakopic, J.; Stampar, F.; Veberic, R.; Jamnik, P. The effect of bioactive compounds on in vitro and in vivo antioxidant activity of different berry juices. *PLoS ONE* **2012**, *7*, e47880. [CrossRef]

21. Reverberi, M.; Zjalic, S.; Ricelli, A.; Punelli, F.; Camera, E.; Fabbri, C.; Picardo, M.; Fanelli, C.; Fabbri, A.A. Modulation of antioxidant defense in Aspergillus parasiticus is involved in aflatoxin biosynthesis: A role for the ApyapA gene. *Eukaryot. Cell* **2008**, *7*, 988–1000. [CrossRef]

22. Roze, L.V.; Hong, S.-Y.; Linz, J.E. Aflatoxin biosynthesis: Current frontiers. *Annu. Rev. Food Sci. Technol.* **2013**, *4*, 293–311. [CrossRef]

23. Roze, L.V.; Laivenieks, M.; Hong, S.-Y.; Wee, J.; Wong, S.S.; Vanos, B.; Awad, D.; Ehrlich, K.C.; Linz, J.E. Aflatoxin biosynthesis is a novel source of reactive oxygen species—A potential redox signal to initiate resistance to oxidative stress? *Toxins* **2015**, *7*, 1411–1430. [CrossRef]

24. Passone, M.A.; Resnik, S.L.; Etcheverry, M.G. In vitro effect of phenolic antioxidants on germination, growth and aflatoxin B_1 accumulation by peanut *Aspergillus* section Flavi. *J. Appl. Microbiol.* **2005**, *99*, 682–691. [CrossRef]

25. Huang, J.-Q.; Jiang, H.-F.; Zhou, Y.-Q.; Lei, Y.; Wang, S.-Y.; Liao, B.-S. Ethylene inhibited aflatoxin biosynthesis is due to oxidative stress alleviation and related to glutathione redox state changes in *Aspergillus flavus*. *Int. J. Food Microbiol.* **2009**, *130*, 17–21. [CrossRef]

26. Galanopoulou, D.; Markaki, P. Study of the Effect of Methyl Jasmonate Concentration on Aflatoxin B(1) Biosynthesis by *Aspergillus parasiticus* in Yeast Extract Sucrose Medium. *Int. J. Microbiol.* **2009**, *2009*, 842626.

27. Degola, F.; Dall'Asta, C.; Restivo, F.M. Development of a simple and high-throughput method for detecting aflatoxins production in culture media. *Lett. Appl. Microbiol.* **2012**, *55*, 82–89. [CrossRef]

28. Reverberi, M.; Fabbri, A.A.; Zjalić, S.; Ricelli, A.; Punelli, F.; Fanelli, C. Antioxidant enzymes stimulation in *Aspergillus parasiticus* by *Lentinula edodes* inhibits aflatoxin production. *Appl. Microbiol. Biotechnol.* **2005**, *69*, 207–215. [CrossRef]

29. Reddy, K.R.N.; Reddy, C.S.; Muralidharan, K. Potential of botanicals and biocontrol agents on growth and aflatoxin production by *Aspergillus flavus* infecting rice grains. *Food Control.* **2009**, *20*, 173–178. [CrossRef]

30. Reddy, K.R.N.; Nurdijati, S.B.; Salle, B. An overview of plant-derived products on control of mycotoxigenic fungi and mycotoxins. *Asian J. Plant Sci.* **2010**, *9*, 126–133. [CrossRef]

31. Cabral, L.D.C.; Pinto, V.F.; Patriarca, A. Application of plant derived compounds to control fungal spoilage and mycotoxin production in foods. *Int. J. Food Microbiol.* **2013**, *166*, 1–14. [CrossRef] [PubMed]

32. Marzouk, B.; Marzouk, Z.; Haloui, E.; Fenina, N.; Bouraoui, A.; Aouni, M. Screening of analgesic and anti-inflammatory activities of *Citrullus colocynthis* from southern Tunisia. *J. Ethnopharmacol.* **2010**, *128*, 15–19. [CrossRef] [PubMed]

33. Gurudeeban, S.; Ramanathan, T. Antidiabetic effect of *Citrullus colocynthis* in alloxon-induced diabetic rats. *Ethno Pharm.* **2010**, *1*, 112–119.

34. Abdel-Hassan, I.A.; Abdel-Barry, J.A.; Tariq Mohammeda, S. The hypoglycaemic and antihyperglycaemic effect of *Citrullus colocynthis* fruit aqueous extract in normal and alloxan diabetic rabbits. *J. Ethnopharmacol.* **2000**, *71*, 325–330. [CrossRef]

35. Mehrzadi, S.; Shojaii, A.; Pur, S.A.; Motevalian, M. Anticonvulsant Activity of Hydroalcoholic Extract of *Citrullus colocynthis* Fruit: Involvement of Benzodiazepine and Opioid Receptors. *J. Evid. Based Complementary Altern. Med.* **2016**, *21*, 31–35. [CrossRef] [PubMed]

36. Hussain, A.I.; Rathore, H.A.; Sattar, M.Z.; Chatha, S.A.; Sarker, S.D.; Gilani, A.H. *Citrullus colocynthis* (L.) Schrad (bitter apple fruit): A review of its phytochemistry, pharmacology, traditional uses and nutritional potential. *J. Ethnopharmacol.* **2014**, *155*, 54–66. [CrossRef]

37. Marzouk, B.; Marzouk, Z.; Décor, R.; Edziri, H.; Haloui, E.; Fenina, N.; Aouni, M. Antibacterial and anticandidal screening of Tunisian *Citrullus colocynthis* Schrad. from Medenine. *J. Ethnopharmacol.* **2009**, *125*, 344–349. [CrossRef]

38. Amine, G.; ELHADJ-Khelil-Aminata, O.; Bouabdallah, G. Evaluation of antifungal effect of organic extracts of Algerian *Citrullus colocynthis* seeds against four strains of Aspergillus isolate from wheat stored. *J. Med. Plants Res.* **2013**, *7*, 727–733.

39. Gowri, S.S.; Priyavardhini, S.; Vasantha, K.; Umadevi, M. Antibacterial activity on *Citrullus colocynthis* Leaf extract. *Anc. Sci. Life* **2009**, *29*, 12–13.

40. Benariba, N.; Djazir, R.; Bellakhdar, W.; Belkacem, N.; Kadiata, M.; Malaisse, W.J.; Sener, A. Phytochemical screening and free radical scavenging activity of *Citrullus colocynthis* seeds extracts. *Asian Pac. J. Trop. Biomed.* **2013**, *3*, 35–40. [CrossRef]

41. Rizvi, T.S.; Mabood, F.; Ali, L.; Al-Broumi, M.; Al Rabani, H.K.M.; Hussain, J.; Jabeen, F.; Manzoor, S.; Al-Harrasi, A. Application of NIR Spectroscopy Coupled with PLS Regression for Quantification of Total Polyphenol Contents from the Fruit and Aerial Parts of *Citrullus colocynthis*. *Phytochem. Anal.* **2018**, *29*, 16–22. [CrossRef]

42. Delazar, A.; Gibbons, S.; Kosari, A.R.; Nazemiyeh, H.; Modarresi, M.; Nahar, L.; Sarker, S.D. Flavone C-glycosides and cucurbitacin glycosides from *Citrullus colocynthis*. *DARU J. Pharm. Sci.* **2006**, *14*, 109–114.

43. Ogbuji, K.; McCutcheon, G.S.; Simmons, A.M.; Snook, M.E.; Harrison, H.F.; Levi, A. Partial Leaf Chemical Profiles of a Desert Watermelon Species (*Citrullus colocynthis*) and Heirloom Watermelon Cultivars (*Citrullus lanatus* var. lanatus). *HortScience* **2012**, *47*, 580–584. [CrossRef]

44. Chawech, R.; Jarraya, R.; Girardi, C.; Vansteelandt, M.; Marti, G.; Nasri, I.; Racaud-Sultan, C.; Fabre, N. Cucurbitacins from the Leaves of *Citrullus colocynthis* (L.) Schrad. *Molecules* **2015**, *20*, 18001–18015. [CrossRef] [PubMed]

45. Hussain, A.I.; Rathore, H.A.; Sattar, M.Z.; Chatha, S.A.; ud din Ahmad, F.; Ahmad, A.; Johns, E.J. Phenolic profile and antioxidant activity of various extracts from *Citrullus colocynthis* (L.) from the Pakistani flora. *Ind. Crop. Prod.* **2013**, *45*, 416–422. [CrossRef]

46. Adam, S.E.I.; Al-Yahya, M.; l-Farhan, A.H. Response of Najdi sheep to oral administration of *Citrullus colocynthis* fruits, Nerium oleander leaves or their mixture. *Small Rumin. Res.* **2001**, *40*, 239–244. [CrossRef]

47. Yoshikawa, M.; Morikawa, T.; Kobayashi, H.; Nakamura, A.; Matsuhira, K.; Nakamura, S.; Matsuda, H. Bioactive saponins and glycosides. XXVII. Structures of new cucurbitane-type triterpene glycosides and antiallergic constituents from *Citrullus colocynthis*. *Chem. Pharm. Bull.* **2007**, *55*, 428–434. [CrossRef]

48. Shawkey, A.M.; Rabeh, M.A.; Abdellatif, A.O. Biofuntional moleculs from *Citrullus colocynthis*: An HPLC/MS analysis in correlation to antimicrobial and anticancer activities. *Adv. Life Sci. Technol.* **2014**, *17*, 51–61.

49. Jian, C.C.; Chiu, M.H.; Nie, R.L.; Cordell, G.A.; Qiu, S.X. Cucurbitacins and cucurbitaneglycosides: Structures and biological activities. *Nat. Prod. Rep.* **2005**, *22*, 386–399.

50. Negri, M.; Salci, T.P.; Shinobu-Mesquita, C.S.; Capoci, I.R.; Svidzinski, T.I.; Kioshima, E.S. Early state research on antifungal natural products. *Molecules* **2014**, *19*, 2925–2956. [CrossRef]

51. Caceres, I.; El Khoury, R.; Medina, Á.; Lippi, Y.; Naylies, C.; Atoui, A.; El Khoury, A.; Oswald, I.; Bailly, J.-D.; Puel, O. Deciphering the anti-aflatoxinogenic properties of eugenol using a large-scale q-PCR approach. *Toxins* **2016**, *8*, 123. [CrossRef]

52. Degola, F.; Bisceglie, F.; Pioli, M.; Palmano, S.; Elviri, L.; Pelosi, G.; Lodi, T.; Restivo, F.M. Structural modification of cuminaldehyde thiosemicarbazone increases inhibition specificity toward aflatoxin biosynthesis and sclerotia development in *Aspergillus flavus*. *Appl. Microbiol. Biotechnol.* **2017**, *101*, 6683–6696. [CrossRef]

53. Abbas, H.K.; Wilkinson, J.; Zablotowicz, R.; Accinelli, C.; Abel, C.; Bruns, H.; Weaver, M. Ecology of *Aspergillus flavus*, regulation of aflatoxin production, and management strategies to reduce aflatoxin contamination of corn. *Toxin Rev.* **2009**, *28*, 142–153. [CrossRef]

54. Ehrlich, K.C.; Moore, G.G.; Mellon, J.E.; Bhatnagar, D. Challenges facing the biological control strategy for eliminating aflatoxin contamination. *World Mycotoxin J.* **2015**, *8*, 225–233. [CrossRef]

55. Bhatnagar, D.; McCormick, S.P. The inhibitory effect of neem (*Azadirachta indica*) leaf extracts on aflatoxin synthesis in *Aspergillus parasiticus*. *J. Am. Oil Chem. Soc.* **1988**, *65*, 1166–1168. [CrossRef]

56. Masood, A.; Ranjan, K.S. The effect of aqueous plant extracts on growth and aflatoxin production by *Aspergillus flavus*. *Lett. Appl. Microbiol.* **1991**, *13*, 32–34. [CrossRef]

57. Ansari, A.A.; Shrivastava, A.K. The effect of eucalyptus oil on growth and aflatoxin production by *Aspergillus flavus*. *Lett. Appl. Microbiol.* **1991**, *13*, 75–77. [CrossRef]

58. Pottier Alapetite, G. Flore de la Tunisie. Angiospermes-dicotylédones, Gamopétales. *Programme Flore Végétation Tunis.* **1981**, 655–1190.

59. Singleton, V.L.; Rossi, J.A. Colorimetry of Total Phenolics with Phosphomolybdic-Phosphotungstic Acid Reagents. *Am. J. Enol. Vitic.* **1965**, *16*, 144–158.

60. Brand-Williams, W.; Cuvelier, M.E.; Berset, C. Use of a free radical method to evaluate antioxidant activity. *LWT-Food Sci. Technol.* **1995**, *28*, 25–30. [CrossRef]

61. Degola, F.; Berni, E.; Restivo, F.M. Laboratory tests for assessing efficacy of atoxigenic *Aspergillus flavus* strains as biocontrol agents. *Int. J. Food Microbiol.* **2011**, *146*, 235–343. [CrossRef] [PubMed]

Article

Phenyllactic Acid Produced by *Geotrichum candidum* Reduces *Fusarium sporotrichioides* and *F. langsethiae* Growth and T-2 Toxin Concentration

Hiba Kawtharani, Selma Pascale Snini, Sorphea Heang, Jalloul Bouajila, Patricia Taillandier, Florence Mathieu * and Sandra Beaufort *

Laboratoire de Génie Chimique, UMR 5503, Université de Toulouse, CNRS, INPT, UPS, 31326 Toulouse, France; hiba.kawtharani@toulouse-inp.fr (H.K.); selma.snini@toulouse-inp.fr (S.P.S.); sorphea.itc@gmail.com (S.H.); jalloul.bouajila@univ-tlse3.fr (J.B.); patricia.taillandier@toulouse-inp.fr (P.T.)
* Correspondence: florence.mathieu@toulouse-inp.fr (F.M.); sandra.beaufort@toulouse-inp.fr (S.B.);
  Tel.: +335-3432-3935 (F.M.); Tel: +335-3432-3746 (S.B.)

Received: 25 February 2020; Accepted: 24 March 2020; Published: 26 March 2020

Abstract: *Fusarium sporotrichioides* and *F. langsethiae* are present in barley crops. Their toxic metabolites, mainly T-2 toxin, affect the quality and safety of raw material and final products such as beer. Therefore, it is crucial to reduce *Fusarium spp.* proliferation and T-2 toxin contamination during the brewing process. The addition of *Geotrichum candidum* has been previously demonstrated to reduce the proliferation of *Fusarium spp.* and the production of toxic metabolites, but the mechanism of action is still not known. Thus, this study focuses on the elucidation of the interaction mechanism between *G. candidum* and *Fusarium spp.* in order to improve this bioprocess. First, over a period of 168 h, the co-culture kinetics showed an almost 90% reduction in T-2 toxin concentration, starting at 24 h. Second, sequential cultures lead to a reduction in *Fusarium* growth and T-2 toxin concentration. Simultaneously, it was demonstrated that *G. candidum* produces phenyllactic acid (PLA) at the early stages of growth, which could potentially be responsible for the reduction in *Fusarium* growth and T-2 toxin concentration. To prove the PLA effect, *F. sporotrichioides* and *F. langsethiae* were cultivated in PLA supplemented medium. The expected results were achieved with 0.3 g/L of PLA. These promising results contribute to a better understanding of the bioprocess, allowing its optimization at an up-scaled industrial level.

Keywords: phenyllactic acid; biocontrol agent; T-2 toxin; *F. langsethiae*; *F. sporotrichioides*; G. candidum; mycotoxin.

Key Contribution: Phenyllactic acid production by *G. candidum* reduces T-2 toxin concentration by reducing *F. langsethiae* and *F. sporotrichioides* growth.

---

1. Introduction

Beer is the most consumed alcoholic beverage worldwide and the third most popular drink overall after water and tea. In 2018, beer production in the European Union was estimated to be nearly 406,050 10^8 L and its consumption was calculated to be around 370,092 10^8 L [1]. Barley is the main ingredient in the brewing process and its quality directly influences the characteristics of the final product. However, barley crops can be contaminated by several fungal species belonging to *Aspergillus*, *Penicillium* and *Fusarium* genera [2]. The latter is the most prevalent genus all over the world and the main genus in Europe [3]. *Fusarium* species are responsible for the production of toxic metabolites called mycotoxins, which are of increasing concern at both health and economic levels [4]. Indeed, recent surveys carried out in Europe have demonstrated that barley crops are frequently contaminated by *Fusarium* species and their associated mycotoxins [5–8]. The use of such contaminated raw materials in the brewing process impacts the quality of the produced beer [9]. *Fusarium* species can produce several kinds of toxins belonging to the trichothecenes family, of which types A and B are commonly found in food and feed. The most important of them are deoxynivalenol (DON), nivalenol (NIV), T-2 and their derivatives: the 15-acetyldeoxynivalenol (15-ADON), 3-acetyldeoxynivalenol (3-ADON) and HT-2 toxin [10].

The T-2 toxin belonging to the type A family was first isolated in *F. tricinctum* cultures, now called *F. sporotrichioides* and then detected in several cereal grains such as wheat, oats, barley and their derivatives. T-2 toxin is mainly produced by *F. sporotrichioides* and *F. langsethiae* [11,12].

T-2 toxin is known to be the most cytotoxic of the type A trichothecenes and has adverse effects on cellular metabolism [13]. It is 1.5–1.7 times more toxic than its deacetylate form HT-2 toxin. Even though its carcinogenicity was proven in certain affected animals, no evidence of such effect was detected in humans. Therefore, the IARC classified the T2-toxin in group 3 as not classifiable with regard to its carcinogenicity for humans [14], thus leading the European Union (EU) to propose recommendations on the presence of T-2 toxin in cereals and cereal products. Thus, the maximum limits in unprocessed cereals are 100 µg/kg for wheat, rye and other cereals, 200 µg/kg for barley (including malting barley) and corn and 1000 µg/kg for oats. [15].

In order to limit mycotoxin contamination, several pre-harvest and/or post-harvest methods can be adopted [16–18]. These techniques either directly target fungal development or limit mycotoxin levels. Pre-harvest methods include good agricultural practices (GAPs) and good manufacturing practices. Crop rotation, tillage and fungicide treatment are mainly implemented to control fungal infection [19,20]. Fungicides are commonly used during agricultural practices but have numerous disadvantages such as detrimental effects on human and animal health, environmental contamination and subsequently, they have a strong impact on microbial biodiversity [21,22]. Indeed, fungicides of the azole family are used in small grain cereals to control *Fusarium spp.* They target the CYP51 (sterol 14α-demethylase) an important enzyme involved in ergosterol biosynthesis, which is essential to maintain fungal membrane fluidity and permeability [23]. By reducing fungal growth, they disturb the natural microbial ecosystem, causing the potential emergence of new microorganisms that may be even more dangerous [24]. Moreover, fungal resistance to these compounds has developed in recent years, thus reducing their effectiveness [25]. In an attempt to limit the proliferation of these toxinogenic and phytopathogens fungal species, biocontrol approaches are starting to be published. Recently, Rahman et al. (2018) proposed the concept of the "plant holobiont". They demonstrated that barley is consistently associated with beneficial bacteria inside their seeds and that this type of association should be encouraged to help the plant react to fungal attack. This could open up new possibilities for applying seeds formulated with endophytic bacteria as bioinoculants for sustainable agriculture [26]. Post-harvest methods include physical treatments such as high temperature treatment exposure and chemical agents. However, these procedures can lead to the deterioration of nutritional quality and alteration of the organoleptic properties of the food matrix [27–29]. Therefore, it is important to conceive a bioprocess to minimize these side effects. This implies the use of natural and environmentally friendly ways to maintain the safety and the quality of the final product. The brewing process comprises several stages and among them, the malting step provides the best conditions (22 °C and high humidity) for *Fusarium* development and T-2 toxin production [30,31]. To reduce mycotoxin concentration during the malting process, several studies have reported the use of lactic acid bacteria (LAB), which are characterized by their antifungal and anti-mycotoxigenic properties [32,33]. However, LAB are fermenting bacteria and can spoil beer, leading to acidification, turbidity, off-flavors and ropiness, depending on the bacterial strain [34,35].

The French Institute for Brewing and Malting (IFBM) filed a patent in September 1999 entitled "The inoculation by *Geotrichum candidum* during malting of cereals or other plants" [36]. The invention consists of using *G. candidum* strain, a filamentous yeast, to inhibit the development of undesirable microorganisms such as *Fusarium spp.* during the malting process to avoid the contamination of beer products by T-2 toxin. Antibacterial activity was previously attributed to this microorganism as it can inhibit the growth of several bacteria such as *Listeria monocytogenes* [37]. *G. candidum* was also found to inhibit other Gram-positive bacteria, such as *Staphylococcus aureus* and *Enterococcus faecalis*, and Gram-negative bacteria, such as *Providencia stuartii* and *Klebsiella oxytoca* [38]. As a matter of fact, three metabolites produced by *G. candidum* have been reported as antimicrobial compounds. Phenyllactic acid (PLA) and indoleacetic acid (ILA) induce behavioral and structural alterations to *L. monocytogenes*, which completely inhibit its growth [37]. The third metabolite, phenylethyl alcohol (PEA), is responsible for the "aromatic rose" character of soft cheese, and promotes membrane damage and inhibition of RNA and protein synthesis of Gram-positive and Gram-negative bacteria, such as *S. aureus* and *Escherichia coli* [39]. Among these three metabolites, PLA is the most effective against bacteria growth [37].

However, the *G. candidum* mechanism against *Fusarium spp.* and T-2 toxin production during the malting process is still unidentified. Given the data in the literature considering PLA as a powerful antimicrobial, the production of PLA by *G. candidum* now needs to be monitored and its effect on *Fusarium spp.* growth as well as on T-2 toxin concentration needs to be quantified.

Thus, this study aims to decipher the interaction mechanisms between *G. candidum* and two *Fusarium* strains: *F. langsethiae* 2297 and *F. sporotrichioides* 186, determine on which level these interactions occur and identify the metabolite responsible for the T-2 toxin concentration reduction.

2. Results

2.1. Effect of Co-Culture between G. candidum and Fusarium Strains on Fungal Growth and T-2 Toxin Concentration

The co-culture experiment consisted of simultaneously inoculating *G. candidum* and *Fusarium* strains into Ym medium for different incubation times (ranging from 24 to 168 h) at 22 °C, 150 rpm. For each incubation time, microbial dry weight, T-2 toxin and PLA concentrations were analyzed in control cultures (*G. candidum*, *F. langsethiae* 2297 and *F. sporotrichioides* 186 alone) and in co-cultures. Two co-culture experiments were conducted: *G. candidum* with *F. langsethiae* 2297 (Gc/Fl) and *G. candidum* with *F. sporotrichioides* 186 (Gc/Fs).

In control cultures, *G. candidum* dry weight increased during the first 3 days of incubation reaching 3.9 g/L and then slightly decreased to stagnate at 2.3 g/L during the last hours of the experiments. For *Fusarium* control cultures, fungal biomass increased throughout the whole experimental duration; *F. langsethiae* 2297 attained a maximum of 3.8 g/L whereas *F. sporotrichioides* 186 almost reached 3 g/L. In co-culture conditions, where microorganisms were simultaneously inoculated, for the two co-culture experiments (Gc/Fl and Gc/Fs), the total biomass increased during the first 3 days of incubation and then stabilized until the end of the experiment. However, in both co-culture experiments, for each incubation time, the total microbial dry weight was not the sum of dry weights obtained separately in control culture. Thus, co-culture leads to microbial growth reduction without distinguishing the growth of *G. candidum* from *Fusarium* species (Figure 1).

Figure 2, Panel A, shows that in control culture (*F. langsethiae* 2297 alone), T-2 toxin was detected from 48 h and the concentration was 99.65 μg/L (±7.28), and reached 332.7 μg/L (±29.42) after an incubation time of 168 h. In co-culture (*G. candidum* with *F. langsethiae* 2297), T-2 toxin was detected from 72 h (20.22 μg/L ± 4.32) and attained 116.44 μg/L (±10.89) after incubation for 168 h. The percentage of T-2 toxin reduction was 100%, 94%, 84% and 65% at 48 h, 72 h, 120 h and 168 h, respectively. These results were similar to those previously obtained for the *F. langsethiae* 033 strain [40]. The same phenomenon was observed in the second co-culture experiment using *F. sporotrichioides* 186 strain with slightly different degrees of reduction (Figure 2, Panel B). In control culture (*F. sporotrichioides* 186 alone), T-2 toxin was detected from 48 h and the concentration was 82.3 μg/L (±6.1), reaching 294.65 μg/L (±4.74) after incubation for 168 h. As for the first co-culture experiment, in the co-culture *G. candidum* with *F. sporotrichioides* 186, T-2 toxin was detected from 72 h (18.8 μg/L ± 6.12) and reached 106.25 μg/L (±3.04) after incubation for 168 h. The percentage of T-2 toxin reduction was 100%, 92%,

74% and 64% at 48 h, 72 h, 120 h and 168 h, respectively. To ensure that T-2 toxin was not degraded, HT-2 toxin was also monitored and was not detected.

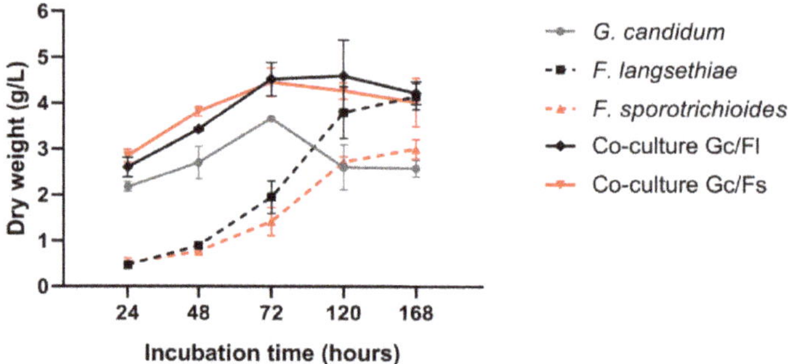

Figure 1. Microbial dry weight analysis in control cultures (*G. candidum*, *F. langsethiae* 2297 and *F. sporotrichioides* 186 alone) and in co-culture experiments (*G. candidum* with *F. langsethiae* 2297 and *G. candidum* with *F. sporotrichioides* 186).

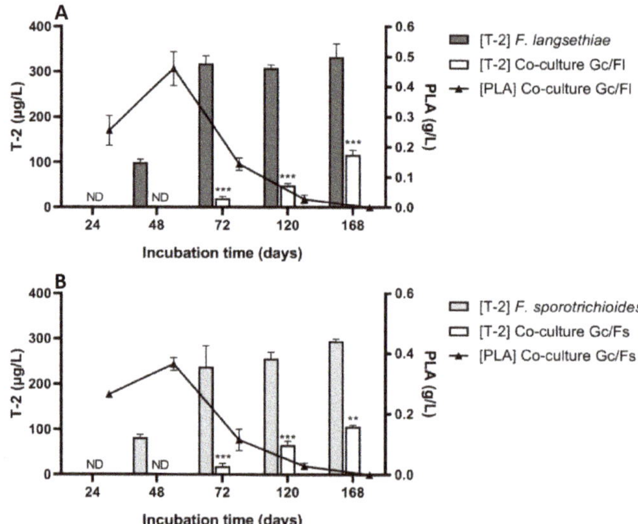

Figure 2. T-2 concentration (µg/L) and phenyllactic acid (PLA) concentration (g/L) in co-culture experiments. Panel **A**: Co-culture experiment of *G. candidum* and *F. langsethiae* 2297. Panel **B**: Co-culture experiment of *G. candidum* and *F. sporotrichioides* 186 (One-way ANOVA, Tukey's multiple comparisons post-hoc test, ** p-value < 0.01; *** p-value < 0.001) ND = not detectable.

In both co-culture experiments, the PLA concentration increased rapidly during the first two days of incubation. In the co-culture with *F. langsethiae* 2297, PLA concentration attained 0.25 g/L (±0.05) at 24 h and 0.46 g/L (±0.06) at 48 h. Afterward, it radically decreased starting at 72 h (0.14 g/L± 0.02) to reach a null value at the end of the incubation time. The same profile was observed in the co-culture with *F. sporotrichioides* 186: PLA concentration attained 0.26 g/L (±0.02) at 24 h and 0.36 g/L (±0.04) at 48 h. PLA concentration in co-culture conditions was inversely proportionate to the T-2 toxin concentration. Indeed, the increase in T-2 toxin concentration was correlated with the reduction of PLA concentration in the medium. When PLA was at its highest level (0.46 g/L in Gc/Fl and 0.36 g/L in Gc/Fs), T-2 toxin was not detected.

2.2. *G. candidum Growth and PLA Production Kinetics*

The *G. candidum* strain selected by the IFBM and used in this study produces PLA during the brewing process. To study the growth of this filamentous yeast, Ym medium was initially inoculated with 0.2 g/L of a *G. candidum* starter culture and incubated at 22 °C, 200 rpm for 5 days. Samples were withdrawn at the starting point and after 6 h, 12 h 24 h, 48 h, 72 h, 96 h, and 120 h of fermentation time and the PLA concentrations were measured.

After 48 h of incubation, the concentration of PLA reached a maximal concentration of 0.41 g/L for 2.25 g/L of yeast dry weight. After 72 h of culture, both *G. candidum* dry weight and PLA concentration started decreasing, growth went from a maximum of 3.43 g/L (±0.51) to 2.46 g/L (±0.46) and PLA concentration drastically decreased from a maximum of 0.41 g/L (±0.03) to 0.03 g/L (±0.01) (almost 17 times less) (Figure 3, Panel A). Figure 3, Panel B demonstrates the specific production of PLA relative to *G. candidum* biomass through the fermentation time. It clearly shows that the PLA is highly accumulated in the medium at the early stages of *G. candidum* growth between 12 and 48 h and then drastically disappears.

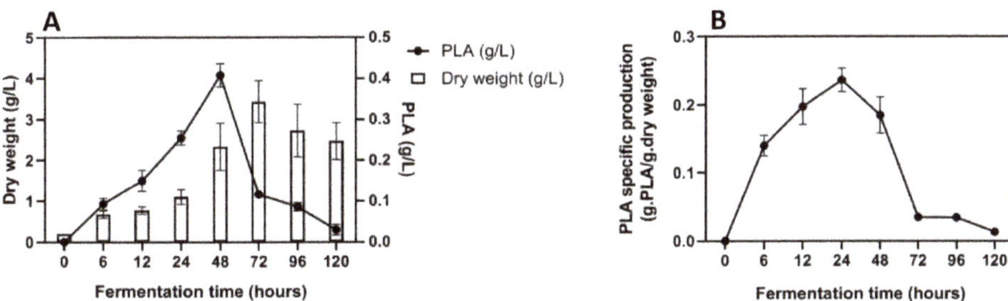

Figure 3. PLA concentration (g/L) and *G. candidum* biomass (g/L) in Ym medium (Panel **A**) and PLA specific production (g PLA/g dry weight) in Ym medium (Panel **B**).

2.3. *Sequential Cultures*

This experiment studied the indirect interactions between *G. candidum* and the two *Fusarium* strains. Therefore, the same Ym medium used in Section 2.2 to grow *G. candidum* was filtrated after 6 h, 12 h, 24 h, 48 h, 72 h, 96 h and 120 h of fermentation time into sterilized Erlenmeyer flasks. Henceforth, the obtained filtrate will be called the "pre-fermented medium" which contains all the metabolites secreted by *G. candidum*.

In the sequential culture experiment with *F. langsethiae* 2297, the dry fungal weight was gradually reduced on pre-fermented media up to 48 h (Figure 4, Panel A). The most significant reduction in the *F. langsethiae* 2297 dry weight occurred in the flasks pre-fermented for 12 h, 24 h and 48 h with a 62%, 72% and 66% reduction percentage, respectively. Beyond 24 h of pre-fermentation, it appeared that *F. langsethiae* 2297 growth increased slowly. In Ym medium pre-fermented for 120 h, the fungal strain was able to proliferate naturally (3.1 g/L of fungal biomass compared to 3.4 g/L in a non-fermented Ym medium). These results demonstrated that the fungal growth inhibition was more efficient in Ym medium pre-fermented for two days by *G. candidum*. Previous results showed that the PLA was produced during the early growth phase of the yeast reaching its peak (between 0.25 and 0.41 g/L of PLA) at around 24–48 h of fermentation time. This suggests that the PLA was involved in the reduction of fungal biomass at a rate of 72% (going from 3.4 g/L in a non-fermented medium to 0.95 g/L in 24 h pre-fermented medium). A significant reduction in T-2 toxin concentration was observed for fungal cultures performed in Ym medium pre-fermented from 6 h to 72 h (Figure 4-Panel B). To ensure that T-2 toxin was not degraded, HT-2 toxin was also monitored and was not detected.

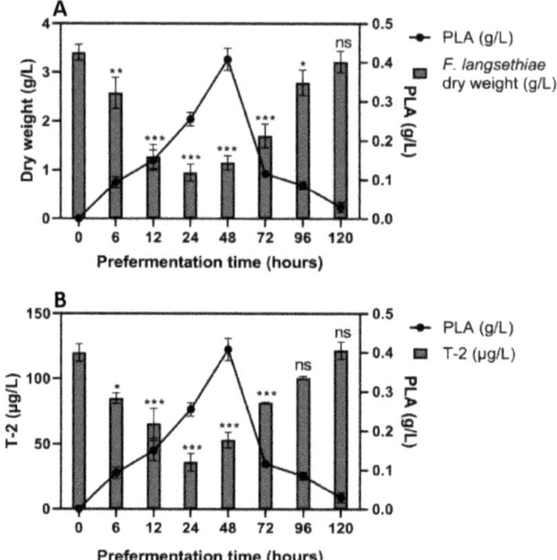

Figure 4. Sequential culture of *F. langsethiae* 2297 inoculated in pre-fermented medium by *G. candidum* and incubated 7 days at 22 °C. Panel **A**: Dry weight of *F. langsethiae* 2297 (g/L) in comparison with PLA concentration (g/L). Panel **B**: T-2 concentration (µg/L) in comparison with PLA concentration (g/L). One-way ANOVA, Dunnett multiple comparisons post-hoc test, * p-value < 0.05; ** p-value < 0.01; *** p-value < 0.001; ns = not significant).

The most significant reduction in T-2 toxin concentration occurred in the flasks pre-fermented for 24 h and 48 h, with a 70% and 56% reduction, respectively. These percentages correlated perfectly with the biomass reduction rate (72% and 66%, respectively). This suggested that the reduction in fungal biomass in the medium is responsible for the reduction in T-2 toxin concentrations. Indeed, specific productions were calculated and demonstrated that the T-2 toxin reduction is correlated to fungal biomass reduction (data not shown).

The same experiment was conducted using *F. sporotrichioides* 186 (Figure 5). Fungal growth was drastically reduced in medium pre-fermented for 6 h, 12 h, 24 h and 48 h at almost the same rate of 70% in correlation with the increase of PLA in the medium. As expected, the Ym medium pre-fermented for 6 h, 12 h, 24 h and 48 h showed an important reduction in T-2 toxin of up to 78%. The equivalence between the growth reduction and the toxin reduction percentages also suggests that it is due to the cessation of fungal growth.

These experiments demonstrated that the interaction between *G. candidum* and *Fusarium* strains occurs through a compound released by *G. candidum* in the medium. As previous results showed, it is highly probable that the PLA, present in *G. candidum* filtrate is the metabolite responsible for the reduction of fungal dry weight and the subsequent reduction in T-2 toxin concentration. To validate this hypothesis, further experiments using pure PLA compound were required.

2.4. Effect of Pure PLA on Fungal Growth and T-2 Toxin Concentration

D-(+)-3Phenyllactic acid was purchased as a pure compound and several concentrations were tested. To validate the results presented in previous sections, PLA solution was prepared at concentrations found at different fermentation times: 0.5 g/L, 0.4 g/L, 0.3 g/L and 0.2 g/L. Lower concentrations of PLA were also tested to determine the minimal inhibitory concentration (MIC): 0.05 g/L and 0.1 g/L. The effect of this pure compound on *Fusarium* strains growth and its ability to produce T-2 toxin was evaluated.

Both *F. langsethiae* 2297 growth and T-2 toxin production were highly affected by the addition of D-PLA in Ym medium (Figure 6). As the concentration of D-PLA increased in the medium, lower fungal mass and lower toxin concentration were quantified. In the control condition (without PLA) *F. langsethiae* 2297 dry weight was 3.2 g/L ($\pm$0.26) and the T-2 toxin concentration was 148 µg/L ($\pm$7.8), whereas in the presence of 0.3 g/L of PLA, both dry weight and T-2 concentration were reduced to a rate of 71%, reaching 0.75 g/L ($\pm$0.7) and 43.4 µg/L ($\pm$1.2), respectively.

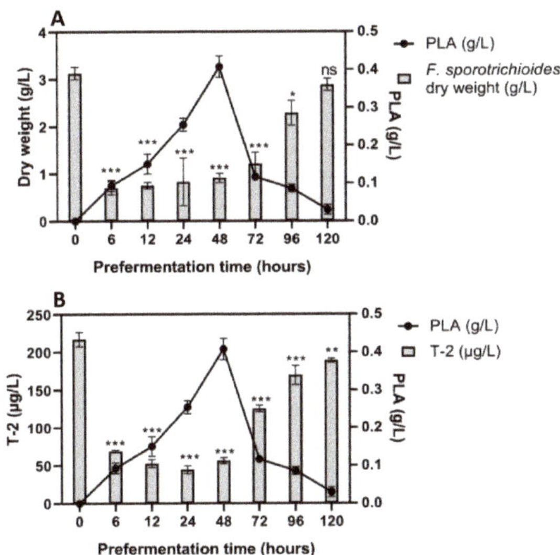

Figure 5. Sequential culture of *F. sporotrichioides* 186 inoculated in pre-fermented medium by *G. candidum* and incubated for 7 days at 22 °C. Panel **A**: Dry weight of *F. sporotrichioides* 186 (g/L) in comparison with PLA concentration (g/L). Panel **B**: T-2 concentration (µg/L) in comparison with PLA concentration (g/L). One-way ANOVA, Dunnett multiple comparisons post-hoc test, * p-value < 0.05; ** p-value < 0.01; *** p-value < 0.001; ns = not significant).

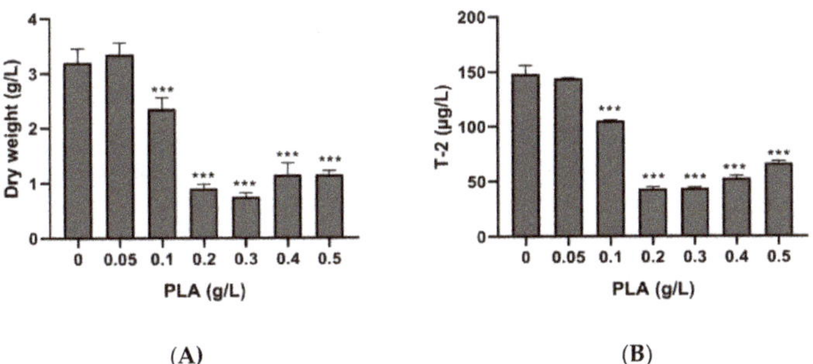

Figure 6. Effect of PLA on the dry weight of *F. langsethiae* 2297 (**A**) and T-2 toxin concentration (**B**) (One-way ANOVA, Dunnett multiple comparisons post-hoc test, *** p-value < 0.001).

The same PLA concentrations were tested on *F. sporotrichioides* 186 and similar results were obtained (Figure 7). The most important reduction occurred in Ym medium supplemented with 0.3 g/L of D-PLA.

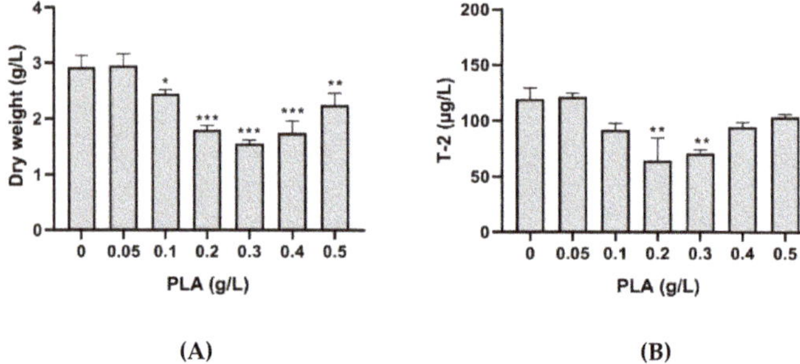

(A) (B)

Figure 7. Effect of phenyllactic acid (PLA) on dry weight of *F. sporotrichioides* 186 (**A**) and T-2 toxin concentration (**B**) (One-way ANOVA, Dunnett multiple comparisons post-hoc test, * *p*-value < 0.05; ** *p*-value < 0.01; *** *p*-value < 0.001).

In both cases, it seems clear that the reduction of T-2 toxin concentration in the medium is directly related to the reduction in fungal growth. Specific production was calculated for each PLA concentration and demonstrated that the T-2 toxin reduction is correlated to the fungal biomass reduction (Figure 8).

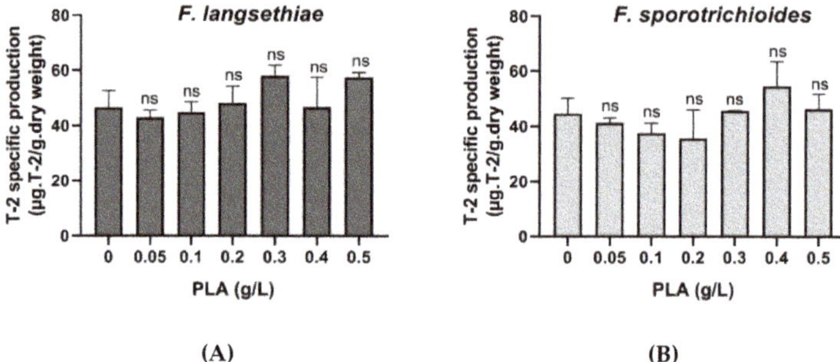

(A) (B)

Figure 8. Specific production of T-2 toxin by *F. langsethiae* 2297 (**A**) and *F. sporotrichioides* 186 in Ym medium supplemented with pure phenyllactic acid (PLA) (**B**) and incubated 7 days at 22 °C (One-way ANOVA, Dunnett multiple comparisons post-hoc test, ns = not significant).

3. Discussion

The contamination of food raw material by fungal species has many consequences. In addition to the alteration of commodities, the loss of nutritional qualities, and the strong reduction in yield, fungal development can lead to the accumulation of toxic compounds such as mycotoxins. In France, the occurrence of several *Fusarium* species in barley crops intended for brewing has become a source of concern over the past ten years. In barley crops, the introduction of *F. sporotrichioides* and *F. langsethiae* has been recently observed, progressively replacing *F. poae* [41–43]. The risk associated with these *Fusarium* species is the production of T-2 toxin, the most toxic compound in the type-A trichothecenes family. During the brewing process, the malting step provides the best conditions (22 °C and high humidity) for *Fusarium* development and T-2 toxin production [30]. Currently, *G. candidum* is used during the brewing process to reduce T-2 toxin contamination. However, its efficiency is variable and the mechanisms of interaction between *G. candidum* and *Fusarium* species are still unknown. Previously, Gastélum-Martinez et al. (2012) used the co-culture method between those two microorganisms and demonstrated that the direct interaction between *G. candidum* and *F. langsethiae* 033 led to a drastic T-2 toxin concentration reduction (93% in comparison to the control culture) [40]. To decipher the mechanism of interaction that lead to this reduction, in the presented study, the two microorganisms were also cultivated sequentially. First, *G. candidum* was cultivated, and removed from the medium before *Fusarium* strain inoculation. Several incubation times for *G. candidum* culture were tested (0 to 146 h) and *Fusarium* incubation was always 7 days. Results obtained in these sequential cultures show a reduction in the T-2 toxin concentration linked to a reduction in fungal growth. In addition, the reduction in T-2 toxin concentration varies according to the medium pre-fermentation time by *G. candidum*. In sequential cultures, T-2 toxin concentrations are inversely correlated with the production of PLA by *G. candidum*, demonstrating that the mechanism leading to T-2 toxin reduction, was linked directly to the PLA concentrations in the medium. Indeed, while the PLA concentration was at its highest level after 48 h of pre-fermentation time, the T-2 toxin concentration was at its lowest. The correlation between *G. candidum* growth evolution and PLA concentration in the medium suggests that PLA is a primary metabolite as it is secreted during the growth phase (from 0 h to 48 h, the PLA concentration varied from 0 to 0.41 g/L) and then gradually disappeared from the culture media. PLA biosynthesis is not yet described in *G. candidum* but well described in lactic acid bacteria (LAB) strains, which can produce large amounts of PLA. In fact, in lactic acid bacteria, the PLA results from amino acid metabolism of phenylalanine and α-ketoglutarate. In a glucose, citric acid or fructose enriched medium, the phenylalanine amino acid group is transferred to α-ketoglutarate under the action of aromatic amino acid transferase (AAT), leading to the formation of phenylpyruvic acid (PPA), an intermediate to PLA. Depending on the type of lactate dehydrogenases (L-LDH or D-LDH) present in lactic acid bacteria, PPA is converted to either L-PLA or D-PLA [44–46]. A potential PLA synthesis pathway is explicitly demonstrated by Chaudhari and Gokhale (2016) and simplified in Figure 9 [47]. Studies have shown that the D form of PLA is more effective as an antimicrobial compound than the L form [37].

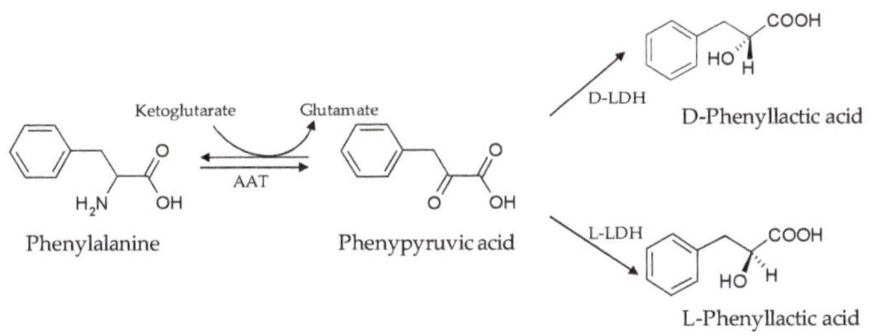

Figure 9. Hypothetical phenyllactic acid biosynthesis pathway. Adapted from Chaudhari and Gokhale (2016) [47]. AAT: amino acid transferase; D-LDH: D-lactate dehydrogenase; L-LDH: L-lactate dehydrogenase.

Several studies have been conducted on LAB and more precisely, on the *Lactobacillus* genus, which is frequently involved in their antifungal activity [48–50]. *Lactobacillus* strains and *L. plantarum* in particular, have been found to produce PLA in sourdough bread. The use of these strains is a means of natural food preservation. Indeed, studies have shown that they improve the shelf life of bread and bakery products by decreasing and/or inhibiting fungal activities. PLA is considered one of the responsible inhibitory compounds along with lactic acid and acetic acid [51]. To our knowledge, no studies have been carried out on PLA metabolism and its toxicity effect in the human body. In 2002, Lavermicocca et al. (2003) studied the fungicidal activity of PLA on 23 fungal strains belonging to *Aspergillus*, *Penicillium* and *Fusarium* genera. Among these strains, 90% showed at least a 50% growth inhibition at PLA concentrations lower than 7.5 g/L. Other strains presented a growth delay of at least three days [52]. Dieuleveux et al. have proved that PLA produced by *G. candidum* strains at 20 g/L also has antibacterial activity against *L. monocytogenes*, *S. aureus*, *E. coli* and *A. hydrophila* [37,38,53].

Fusarium strains used in this study were more susceptible to PLA that those tested by Lavermicocca et al. (2003). Indeed, *F. langsethiae* 2297 growth was drastically reduced (72%) when it was exposed to 0.2 g/L of PLA, whereas *F. sporotrichioides* 186 growth was slightly reduced (47%) when it was exposed to 0.3 g/L of PLA. Thus, there is relevant variability in susceptibility among fungal species. Although the antimicrobial action mechanism is still not elucidated, some suggest that the PLA causes the bacteria to form aggregates with the secretion of polysaccharides described as a "response to the attack". Indeed, as the concentration of PLA increased, a larger amount of polysaccharides were found in the medium and alteration in cell wall rigidity after only 27 h of incubation was observed leading to cell death [38,47]. In this study, as indicated by the specific production of T-2 toxin obtained for each fungal strain, the reduction in the concentration of T-2 toxin is correlated with the reduction in fungal growth. However, in some cases, the inhibition of fungal growth by sub-lethal concentrations of fungicide or some natural products enhances mycotoxin production [54–56]. This must be taken into account in the development of biocontrol strategies.

In this study, to provide an explanation for the phenomenon of T-2 toxin concentration control during the malting process previously observed by the IFBM, in vitro experiments were carried out under environmental conditions close to those of the brewing process. Currently, the filamentous yeast is added in a freeze-dried form (100 g per 25 tons of barley) directly into the barley steeping water for at least 10 h. Then, the water is discarded and the steeped barley remains at rest for 3 to 5 days at 16–20 °C. This stage is the most critical step in the brewing process because the operating conditions favor *Fusarium* growth and T-2 toxin production. Results demonstrate that the reduction in *Fusarium* contamination and T-2 toxin during the malting process is due to the PLA produced by *G. candidum*. Based on the results of this study, in order to develop an effective biocontrol method to use *G. candidum*, preparation of the strain seems essential to activate the PLA production metabolism. Mu et al. developed a medium favorable to PLA production by *Lactobacillus sp.* strains, highly enriched with glucose, phenylpyruvic acid (phenylalanine intermediate in the PLA biosynthesis pathway) and yeast extract [46]. This medium significantly enhanced *Lactobacillus sp.* proliferation, and thus PLA yield. However, the use of such broth on an industrial level does not seem to be applicable for several reasons. On one hand, it may alter the organoleptic characteristics of the final product. On the other, using these components in large amounts would have a considerable economic impact on the industry. Consequently, it seems important to combine optimized growth factors (*G. candidum* activation medium and initial concentration, fermentation duration, temperature, water activity, rotation speed, oxygenation levels, etc.) to enhance PLA production naturally, and to develop an ecofriendly, toxin-free beer product. Moreover, the presence of PLA during the malting step not only helps to reduce *Fusarium* flora and consequently, to reduce T-2 toxin concentration, but it also improves the organoleptic properties of the final beer product [36,57,58]. This study demonstrates for the first time, the role of PLA as a biocontrol agent in reducing T-2 toxin concentration.

4. Materials and Methods

4.1. Reagents and Chemicals

T-2 toxin and phenyllactic acid (PLA) were purchased from Sigma-Aldrich (Saint-Quentin-Fallavier, France). Stock solutions were prepared in dimethylsulfoxyd (DMSO) and acetonitrile–water (30:70 v/v) mixture, respectively, and stored at −18 °C until use. Solvents used for T-2 toxin extraction and high-performance liquid chromatography (HPLC) were analytical grade quality and purchased from Thermo-Fisher Scientific (Illkirch, France). Ultrapure water used for HPLC was purified at 0.22 μm by an ELGA purification system (ELGA LabWater, High Wycombe, United Kingdom).

4.2. Strains, Media and Culture Conditions

In this study, two *Fusarium* strains were used: *F. sporotrichioides* 186 and *F. langsethiae* 2297. Both strains were previously isolated from contaminated barley kernels and were kindly provided by the French Institute of Brewing and Malting (IFBM). The filamentous yeast *Geotrichum candidum* is already used as a biocontrol agent during the malting process (IFBM Malting Yeast®, DMS food specialties, La Ferté sous Jouarre, France) and was purchased from DSM Food Specialties.

Fusarium pre-cultures were performed on potato dextrose agar medium (PDA 39 g/L) and incubated at 22 °C for 7 days. Cultures were then used to induce sporulation or conserved at 4 °C. *Fusarium* strains sporulation was induced in carboxymethylcellulose (CMC) liquid medium (CMC: carboxymethylcellulose 15 g/L; yeast extract 1 g/L; $MgSO_4$ $7H_2O$ 0.5 g/L; NH_4NO_3 1g/L; KH_2PO_4 1 g/L). Briefly, at least 15 plugs of each *Fusarium* strain from a seven-day-old solid pre-culture were inoculated in 150 mL of CMC medium and incubated in an orbital shaker set at 22 °C at 150 rpm for 15 days in the dark. At the end of the incubation time, the solution was filtrated using sterilized Mira cloth. Spores were counted on Thoma cell counting chamber and ultimately used to inoculate culture during further experiments or conserved in 40% glycerol at −80 °C.

G. *candidum* strain was supplied in freeze-dried form, thus a pre-culture was essential to revivify it prior to experimental use. A 24 g/L culture was prepared in 250 mL of yeast and malt (Ym) liquid medium (Ym: glucose 5 g/L; yeast extract 1.5 g/L; malt extract 1.5 g/L; peptone salt 2.5 g/L pH 7) and incubated in an orbital shaker set at 22 °C at 150 rpm for 24 h. At the end of the incubation time, this culture was used as a starter culture.

Ym liquid medium was used during all experiments (co-cultures and sequential cultures) to elucidate the interaction mechanisms between G. *candidum* and *Fusarium* strains.

4.3. Kinetic of PLA Production by G. candidum

In an Erlenmeyer flask, 150 mL of Ym medium was inoculated with G. *candidum* starter culture with a final concentration adjusted at 0.2 g/L and then incubated in an orbital shaker set at 22 °C at 150 rpm for different fermentation times ranging from 6 h to 120 h. At the end of the fermentation time, the medium was aseptically divided into two volumes. First, 50 mL were used to evaluate G. *candidum* growth by measuring the dry weight and PLA concentration by HPLC-DAD at each sampling time. The remaining 100 mL was aseptically filtered to eliminate G. *candidum* cells, leaving only its excreted metabolites in the medium. The medium nutrients were then adjusted according to the volume and the pH was adjusted at 7. These volumes were used during the sequential cultures experiments and are henceforth referred to as pre-fermented medium. Experiments were conducted four times in triplicate.

4.4. Co-Culture of Fusarium Strains and G.candidum

Erlenmeyer flasks containing 150 mL of Ym medium were inoculated with G. *candidum* starter culture at the final concentration of 0.2 g/L. Then, F. *langsethiae* 2297 or F. *sporotrichioides* 186 was inoculated at a final concentration of 10^6 spores/mL in their respective flasks. For control conditions, each microorganism was inoculated alone at the same concentrations. Cultures were incubated in an orbital shaker set at 22 °C at 150 rpm. Several incubation times were tested: 24 h, 48 h, 72 h, 120 h and 168 h. At the end of all sampling times for all culture conditions the total dry weight, PLA and T-2 toxin concentration were evaluated. All experiments were conducted twice in duplicate.

4.5. Sequential Cultures of Fusarium Strains and G.candidum

For sequential cultures, 100 mL of pre-fermented Ym medium at different fermentation times ranging from 6 h to 120 h (used in Section 4.3) were inoculated with F. *langsethiae* 2297 or F. *sporotrichioides* 186 at the final concentration of 10^6 spores/mL in their respective flasks. Cultures were incubated in an orbital shaker set at 22 °C at 150 rpm for 7 days. For the control condition, F. *sporotrichioides* 186 or F. *langsethiae* 2297 were inoculated in a non-fermented Ym liquid medium at the same concentrations. At the end of the incubation time, fungal growth was evaluated by measuring the dry weight and T-2 toxin concentration by HPLC-DAD. All experiments were conducted twice in duplicates.

4.6. Phenyllactic Acid Effect on F. sporotrichioides 186 and F. langsethiae 2297 Growth and T-2 Toxin Concentration

To confirm that PLA is the metabolite produced by G. *candidum*, which is involved in *Fusarium* growth reduction and T-2 toxin concentration reduction, fungal cultures were conducted in Ym liquid medium supplemented with PLA. PLA standard stock solution was prepared at 40 mg/mL in a mixture of acetonitrile/water (30/70, v/v) and appropriate volumes of PLA stock solution were added in order to obtain several different concentrations: 0.05 g/L; 0.1 g/L; 0.2 g/L; 0.3 g/L; 0.4 g/L and 0.5 g/L in Erlenmeyer flasks containing 100 mL of Ym liquid medium. Then, *F. langsethiae* 2297 or *F. sporotrichioides* 186 was inoculated at the final concentration of 10^6 spores/mL in their respective flasks. Cultures were incubated in an orbital shaker set at 22 °C at 150 rpm for 7 days. At the end of the incubation time, *Fusarium* strains' growth was evaluated by measuring the dry weight and T-2 toxin concentration by HPLC-DAD. PLA dilutions were prepared to add only 75 μL of acetonitrile in the culture medium, this concentration having been identified as a no-effect dose on both fungal growth and T-2 toxin concentration. Control cultures were performed by adding only 75 μL of acetonitrile to the medium.

4.7. G. candidum, F. sporotrichioides 186 and F. langsethiae 2297 Biomass Evaluation

To estimate microorganism growth during the incubation period, vacuum filtration was performed to determine the dry weight (g/L). First, cellulose nitrate filters (pore size 0.45 μm, Sartorius Stedim Biotech, Goettingen, Germany) were left to dry overnight in an oven set at 105 °C. Afterward, 10 mL of culture medium were vacuum-filtered at each sampling time and filters were then incubated at 105 °C for 24 h. The microorganism dry weight refers to the difference between filters post-filtration and pre-filtration.

4.8. PLA and T-2 Toxin Quantification by HPLC-DAD

4.8.1. PLA Quantification

At each sampling time, 1 mL of culture media was withdrawn and filtrated through 0.45 μm PTFE syringe filters (Thermo Scientific Fisher, Villebon-Sur-Yvette, France) to eliminate microorganisms from the supernatant prior to injection into HPLC apparatus. Analyses of PLA were performed using a Luna C18(2) column (5 μm, 250 × 4.6 mm) and a pre-column with the same characteristics (Phenomenex, Torrance, CA, USA). The detection of PLA was performed using a Dionex Ultimate 3000 UHPLC system coupled with a diode-array detector (DAD) set at 210 nm (Thermo Fisher Scientific, Illkirch, France). The analysis was performed in a gradient mode using acidified water (0.2% of acetic acid glacial) as solvent A and pure HPLC grade acetonitrile as solvent B. Flow was set at 1.2 mL/min with A/B ratios of 90:10, 50:50, 50:50, 0:100 and 90:10, with run times of 0.0, 4.0, 9.0, 10.0 and 15.0 min, respectively. Injection volume was set at 50 μL. PLA quantification was calculated according to a standard calibration curve with concentrations ranging between 10 and 1000 mg/L.

4.8.2. T-2 Toxin Extraction and Quantification

After the incubation period, cultures were filtrated with Nalgene™ Rapid-Flow™ Filters of 0.45 μm pore size (Thermofischer Scientific, Waltham, MA, USA) to remove microorganisms. Filtrates were then extracted with 70 mL of ethyl acetate and shaken on a Universal Shaker SM 30 B Control Edmund Bühler® (Thermofischer Scientific, Waltham, MA, USA) set at 150 rpm overnight. The organic phase was recovered and evaporated until dry under a rotavapor set at 60 °C. Samples were resuspended with 2 mL of acetonitrile/water (30/70, v/v) mixture and filtered through 0.45 μm PTFE syringe filters (Sigma Aldrich, St. Quentin Fallavier, France). Samples were conserved at 4 °C until further analysis. T-2 toxin was analyzed by Gemini C18 columns, 150 mm × 4.6 mm, 3 μm and a pre-column with the same characteristics (Phenomenex). As for PLA, T-2 toxin was detected and quantified using HPLC-DAD (Dionex, Sunnyvale, CA, USA) according to the methodology described by Medina et al. [59]. T-2 toxin quantification was calculated according to a standard calibration curve with concentrations ranging between 0.2 and 50 μg/mL.

4.9. Statistical Analysis

First, the normal distribution of data was tested by the Shapiro-Wilk test. Then, one-way analysis of variance (ANOVA) followed by Dunnett's multiple comparisons test was used to analyze the effect of PLA on *F. langsethiae 2297* and *F. sporotrichioides 186* growth and their T-2 production. One-way ANOVA followed by a Tukey's multiple comparisons test was used to analyze the differences between control and co-culture or sequential culture conditions. The statistical analysis of data was carried out with GraphPad Prism 8 software (GraphPad Software, La Jolla, CA, USA). Differences were considered to be statistically significant when the *p*-value was lower than 0.05. Graphical values are represented by mean ± standard deviation (SD).

Author Contributions: Conceptualization, S.P.S., P.T., F.M. and S.B.; Formal analysis, H.K. and S.P.S.; Funding acquisition, S.B.; Investigation, H.K. and S.H.; Methodology, J.B., P.T. and F.M.; Supervision, S.P.S., P.T., F.M. and S.B.; Writing—original draft, H.K.; Writing—review & editing, S.P.S., P.T., F.M. and S.B. All authors have read and agreed to the published version of the manuscript.

Funding: This research was funded by the Agence Nationale de la Recherche (ANR-16-CE21-0011).

Acknowledgments: The authors would like to thank the French Institute of Brewing and Malting for providing microbial strains. The authors would also like to thank Philippe Anson for his technical support.

Conflicts of Interest: The authors declare no conflict of interest.

References

1. The Brewers of Europe. *European Beer Trends—Statistics Report*; The Brewers of Europe: Brussel, Belgium, 2019.
2. Laitila, A. Toxigenic fungi and mycotoxins in the barley-to-beer chain. In *Food Science, Technology and Nutrition, Brewing Microbiology*; Woodhead Publishing Series; Hill, E.A., Ed.; Woodhead Publishing: Oxford, UK, 2015; pp. 107–139. ISBN 978-1-78242-331-7.
3. Creppy, E.E. Update of survey, regulation and toxic effects of mycotoxins in Europe. *Toxicol. Lett.* **2002**, *127*, 19–28. [CrossRef]
4. Bennett, J.W.; Klich, M. Mycotoxins. *Clin. Microbiol. Rev.* **2013**, *16*, 497–516. [CrossRef]
5. Kirinčič, S.; S ̌krjanc, B.; Kos, N.; Kozolc, B.; Pirnat, N.; Tavčar-Kalcher, G. Mycotoxins in cereals and cereal products in Slovenia—Official control of foods in the years 2008–2012. *Food Control* **2015**, *50*, 157–165. [CrossRef]
6. Pleadin, J.; Vahčić, N.; Perši, N.; Ševelj, D.; Markov, K.; Frece, J. *Fusarium* mycotoxins' occurrence in cereals harvested from Croatian fields. *Food Control* **2013**, *32*, 49–54. [CrossRef]
7. Běláková, S.; Benešová, K.; Čáslavský, J.; Svoboda, Z.; Mikulíková, R. The occurrence of the selected *fusarium* mycotoxins in czech malting barley. *Food Control* **2014**, *37*, 93–98. [CrossRef]
8. Morcia, C.; Tumino, G.; Ghizzoni, R.; Badeck, F.W.; Lattanzio, V.M.T.; Pascale, M.; Terzi, V. Occurrence of *Fusarium langsethiae* and T-2 and HT-2 toxins in Italian malting barley. *Toxins (Basel)* **2016**, *8*, 247. [CrossRef] [PubMed]
9. Pascari, X.; Ramos, A.J.; Marín, S.; Sanchís, V. Mycotoxins and beer. Impact of beer production process on mycotoxin contamination. A review. *Food Res. Int.* **2018**, *103*, 121–129. [CrossRef] [PubMed]
10. Donnell, K.O.; Mccormick, S.P.; Busman, M.; Proctor, R.H.; Ward, J.; Doehring, G.; Geiser, D.M.; Alberts, J.F.; Rheeder, J.P.; Donnell, K.O.; et al. 1984 "Toxigenic *Fusarium* Species: Identity and Mycotoxicology" revisited. *Mycologia* **2018**, *110*, 1058–1080. [CrossRef]
11. Gilgan, M.W.; Smalley, E.B.; Strong, F.M. Isolation and partial characterization of a toxin from *Fusarium tricinctum* on moldy corn. *Arch. Biochem. Biophys.* **1966**, *114*, 1–3. [CrossRef]
12. Thrane, U.; Adler, A.; Clasen, P.E.; Galvano, F.; Langseth, W.; Lew, H.; Logrieco, A.; Nielsen, K.F.; Ritieni, A. Diversity in metabolite production by *Fusarium langsethiae, Fusarium poae*, and *Fusarium sporotrichioides*. *Int. J. Food Microbiol.* **2004**, *95*, 257–266. [CrossRef] [PubMed]
13. van der Fels-Klerx, H.; Stratakou, I. T-2 toxin and HT-2 toxin in grain and grain-based commodities in Europe: Occurrence, factors affecting occurrence, co-occurrence and toxicological effects. *World Mycotoxin J.* **2010**, *3*, 349–367. [CrossRef]
14. IARC. Some naturally occurring substances: Food items and constituents, heterocyclic aromatic amines and mycotoxins. *IARC Monogr. Eval. Carcinog. Risks Humans.* **1993**, *56*, 245–395.
15. European Commission (EC). Recomendations on the presence of T-2 and HT-2 toxin in cereals and cereal products. *Off. J. Eur. Union* **2013**, *56*, 12–15. [CrossRef]

16. Fandohan, P.; Gnonlonfin, B.; Hell, K.; Marasas, W.F.O.; Wingfield, M.J. Natural occurrence of *Fusarium* and subsequent fumonisin contamination in preharvest and stored maize in Benin, West Africa. *Int. J. Food Microbiol.* **2005**, *99*, 173–183. [CrossRef] [PubMed]

17. Adegoke, G.O.; Letuma, P. Strategies for the Prevention and Reduction of Mycotoxins in Developing Countries. In *Mycotoxin and Food Safety in Developing Countries*; Makun, H., Ed.; IntechPublisher: Rijeka, Croatia, 2013; pp. 123–136.

18. Magan, N.; Aldred, D. Post-harvest control strategies: Minimizing mycotoxins in the food chain. *Int. J. Food Microbiol.* **2007**, *119*, 131–139. [CrossRef] [PubMed]

19. Agriopoulou, S.; Stamatelopoulou, E.; Varzakas, T. Control Strategies: Prevention and Detoxification in Foods. *Foods* **2020**, *9*, 137. [CrossRef]

20. Ferrigo, D.; Raiola, A.; Causin, R. Fusarium Toxins in Cereals: Occurrence, Legislation, Their Management. *Molecules* **2016**, *21*, 627. [CrossRef]

21. Gupta, P.K.; Aggarwal, M. Toxicity of fungicides. In *Veterinary Toxicology*, 3rd ed.; Academic Press: Cambridge, MA, USA, 2018; pp. 569–580. ISBN 978-0-12-385926-6.

22. Zubrod, J.; Bundschuh, M.; Arts, G.; Brühl, C.; Imfeld, G.; Knäbel, A.; Payraudeau, S.; Rasmussen, J.; Rohr, J.; Scharmüller, A.; et al. Fungicides—An Overlooked Pesticide Class ? *Env. Sci Technol* **2019**, *53*, 3347–3365. [CrossRef]

23. Mansfield, B.E.; Oltean, H.N.; Oliver, B.G.; Hoot, S.J.; Leyde, S.E.; Hedstrom, L.; White, T.C. Azole drugs are imported by facilitated diffusion in *Candida albicans* and other pathogenic fungi. *PLoS Pathog.* **2010**, *6*, 1–11. [CrossRef]

24. Tano, Z.J. Ecological Effects of Pesticides. In *Pesticides in the Modern Word—Risk and Benefits*; Stoytcheva, M., Ed.; IntechPublisher: Rijeka, Croatia, 2011; pp. 533–540.

25. Price, C.L.; Parker, J.E.; Warrilow, A.G.; Kelly, D.E.; Kelly, S.L. Azole fungicides—understanding resistance mechanisms in agricultural fungal pathogens. *Pest Manag. Sci.* **2015**, *71*, 1054–1058. [CrossRef]

26. Rahman, M.M.; Flory, E.; Koyro, H.W.; Abideen, Z.; Schikora, A.; Suarez, C.; Schnell, S.; Cardinale, M. Consistent associations with beneficial bacteria in the seed endosphere of barley (*Hordeum vulgare* L.). *Syst. Appl. Microbiol.* **2018**, *41*, 386–398. [CrossRef] [PubMed]

27. Jackson, L.S.; Katta, S.K.; Fingerhut, D.D.; DeVries, J.W.; Bullerman, L.B. Effects of Baking and Frying on the Fumonisin B 1 Content of Corn-Based Foods. *J. Agric. Food Chem.* **2002**, *45*, 4800–4805. [CrossRef]

28. Kabak, B.; Dobson, A.D.W.; Var, I. Strategies to prevent mycotoxin contamination of food and animal feed: A review. *Crit. Rev. Food Sci. Nutr.* **2006**, *46*, 593–619. [CrossRef] [PubMed]

29. Terzi, V.; Tumino, G.; Stanca, A.M.; Morcia, C. Reducing the incidence of cereal head infection and mycotoxins in small grain cereal species. *J. Cereal Sci.* **2014**, *59*, 284–293. [CrossRef]

30. Strub, C.; Pocaznoi, D.; Lebrihi, A.; Fournier, R.; Mathieu, F. Influence of barley malting operating parameters on T-2 and HT-2 toxinogenesis of *Fusarium langsethiae*, a worrying contaminant of malting barley in Europe. *Food Addit. Contam. Part A Chem. Anal. Control. Expo. Risk Assess.* **2010**, *27*, 1247–1252. [CrossRef]

31. Mastanjevi, K.; Krstanovic, V.; Mastanjevic, K.; Šarkanj, B. Malting and Brewing Industries Encounter *Fusarium spp.* Related Problems. *Toxins (Basel)* **2018**, *4*, 3. [CrossRef]

32. Sadiq, F.A.; Yan, B.; Tian, F.; Zhao, J.; Zhang, H.; Chen, W. Lactic Acid Bacteria as Antifungal and Anti-Mycotoxigenic Agents: A Comprehensive Review. *Compr. Rev. food Sci. food Saf.* **2019**, *18*. [CrossRef]

33. Rouse, S.; van Sinderen, D. Bioprotective Potential of Lactic Acid Bacteria in Malting and Brewing. *J. Food Prot.* **2008**, *71*, 1724–1733. [CrossRef]

34. Geissler, A.J.; Behr, J.; Kamp, K. Von Vogel, R.F. Metabolic strategies of beer spoilage lactic acid bacteria in beer. *Int. J. Food Microbiol.* **2016**, *216*, 60–68. [CrossRef]

35. Suzuki, K. 125th Anniversary Review: Microbiological Instability of Beer Caused by Spoilage Bacteria. *J. Inst. Brew.* **2011**, *117*, 131–155. [CrossRef]

36. Boivin, P.; Malanda, M. Inoculation by Geotrichum Candidum during Malting of Cereals or Other Plants. US Patent 5,955,070, 21 September 1999.

37. Dieuleveux, V.; Van Der Pyl, D.; Chataud, J.; Gueguen, M. Purification and characterization of anti-Listeria compounds produced by *Geotrichum candidum*. *Appl. Environ. Microbiol.* **1998**, *64*, 800–803. [CrossRef]

38. Dieuleveux, V.; Lemarinier, S.; Guéguen, M. Antimicrobial spectrum and target site of D-3-phenyllactic acid. *Int. J. Food Microbiol.* **1998**, *40*, 177–183. [CrossRef]

39. Lucchini, J.J.; Corre, J.; Cremieux, A. Antibacterial activity of phenolic compounds and aromatic alcohol. *Res. Microbiol.* **1990**, *141*, 499–510. [CrossRef]

40. Gastélum-Martínez, E.; Compant, S.; Taillandier, P.; Mathieu, F. Control of T-2 toxin in *Fusarium langsethiae* and *Geotrichum candidum* co-culture. *Arh. Hig. Rada Toksikol.* **2012**, *63*, 447–456. [CrossRef] [PubMed]

41. Torp, M.; Nirenberg, H.I. *Fusarium langsethiae sp. nov.* on cereals in Europe. *Int. J. Food Microbiol.* **2004**, *95*, 247–256. [CrossRef] [PubMed]

42. Imathiu, S.M.; Edwards, S.G.; Ray, R.V.; Back, M.A. *Fusarium langsethiae*—A HT-2 and T-2 Toxins Producer that Needs More Attention. *J. Phytopathol.* **2013**, *161*, 1–10. [CrossRef]

43. Foroud, N.A.; Baines, D.; Gagkaeva, T.Y.; Thakor, N.; Badea, A.; Steiner, B.; Bürstmayr, M.; Bürstmayr, H. Trichothecenes in Cereal Grains – An Update. *Toxins (Basel)* **2019**, *11*, 634. [CrossRef]

44. Vermeulen, N.; Gánzle, M.G.; Vogel, R.F. Influence of peptide supply and cosubstrates on phenylalanine metabolism of *Lactobacillus sanfranciscensis* DSM20451T and *Lactobacillus plantarum* TMW1.468. *J. Agric. Food Chem.* **2006**, *54*, 3832–3839. [CrossRef]

45. Li, X.; Jiang, B.; Pan, B. Biotransformation of phenylpyruvic acid to phenyllactic acid by growing and resting cells of a *Lactobacillus* sp. *Biotechnol. Lett.* **2007**, *29*, 593–597. [CrossRef]

46. Mu, W.; Chen, C.; Li, X.; Zhang, T.; Jiang, B. Optimization of culture medium for the production of phenyllactic acid by *Lactobacillus* sp. SK007. *Bioresour. Technol.* **2009**, *100*, 1366–1370. [CrossRef]

47. Chaudhari, S.; Gokhale, D. Phenyllactic Acid: A Potential Antimicrobial Compound in Lactic acid Bacteria. *J. Bacteriol. Mycol. Open Access* **2016**, *2*, 121–125. [CrossRef]

48. Corsetti, A.; Gobbetti, M.; Rossi, J.; Damiani, P. Antimould activity of sourdough lactic acid bacteria: Identification of a mixture of organic acids produced by *Lactobacillus sanfrancisco* CB1. *Appl. Microbiol. Biotechnol.* **1998**, *50*, 253–256. [CrossRef] [PubMed]

49. Hassan, Y.I.; Bullerman, L.B. Antifungal activity of *Lactobacillus paracasei ssp. tolerans* isolated from a sourdough bread culture. *Int. J. Food Microbiol.* **2008**, *121*, 112–115. [CrossRef] [PubMed]

50. Sathe, S.J.; Nawani, N.N.; Dhakephalkar, P.K.; Kapadnis, B.P. Antifungal lactic acid bacteria with potential to prolong shelf-life of fresh vegetables. *J. Appl. Microbiol.* **2007**, *103*, 2622–2628. [CrossRef]

51. Lavermicocca, P.; Valerio, F.; Evidente, A.; Lazzaroni, S.; Corsetti, A.; Gobbetti, M. Purification and characterization of novel antifungal compounds from the sourdough *Lactobacillus plantarum* strain 21B. *Appl. Environ. Microbiol.* **2000**, *66*, 4084–4090. [CrossRef]

52. Lavermicocca, P.; Valerio, F.; Visconti, A. Antifungal activity of phenyllactic acid against molds isolated from bakery products. *Appl. Environ. Microbiol.* **2003**, *69*, 634–640. [CrossRef]

53. Dieuleveux, V.; Guéguen, M. Antimicrobial effects of D-3-phenyllactic acid on *Listeria monocytogenes* in TSB-YE medium, milk, and cheese. *J. Food Prot.* **1998**, *61*, 1281–1285. [CrossRef]

54. Morcia, C.; Tumino, G.; Ghizzoni, R.; Bara, A.; Salhi, N.; Terzi, V. In Vitro Evaluation of Sub-Lethal Concentrations of Plant-Derived Antifungal Compounds on FUSARIA Growth and Mycotoxin Production. *Molecules* **2017**, *22*, 1271. [CrossRef]

55. Mateo, E.M.; Gómez, J.V.; Gimeno-Adelantado, J.V.; Romera, D.; Mateo-Castro, R.; Jiménez, M. Assessment of azole fungicides as a tool to control growth of *Aspergillus flavus* and aflatoxin B1 and B2 production in maize. *Food Addit. Contam. Part A Chem. Anal. Control. Expo. Risk Assess.* **2017**, *34*, 1039–1051. [CrossRef]

56. Audenaert, K.; Vanheule, A.; Höfte, M.; Haesaert, G. Deoxynivalenol: A Major Player in the Multifaceted Response of *Fusarium* to Its Environment. *Toxins (Basel).* **2013**, *6*, 1. [CrossRef]

57. Piegza, M.; Witkowska, D.; Stempniewicz, R. Enzymatic and molecular characteristics of *Geotrichum candidum* strains as a starter culture for malting. *J. Inst. Brew.* **2014**, *120*, 341–346. [CrossRef]

58. Hattingh, M.; Alexander, A.; Meijering, I.; van Reenen, C.A.; Dicks, L.M.T. Malting of barley with combinations of *Lactobacillus plantarum, Aspergillus niger, Trichoderma reesei, Rhizopus oligosporus* and *Geotrichum candidum* to enhance malt quality. *Int. J. Food Microbiol.* **2014**, *173*, 36–40. [CrossRef] [PubMed]

59. Medina, A.; Valle-Algarra, F.M.; Jiménez, M.; Magan, N. Different sample treatment approaches for the analysis of T-2 and HT-2 toxins from oats-based media. *J. Chromatogr. B Anal. Technol. Biomed. Life Sci.* **2010**, *878*, 2145–2149. [CrossRef] [PubMed]

Article

Fullerol $C_{60}(OH)_{24}$ Nanoparticles Affect Secondary Metabolite Profile of Important Foodborne Mycotoxigenic Fungi In Vitro

Tihomir Kovač [1,2,*,†], Bojan Šarkanj [1,2,3,†], Ivana Borišev [4], Aleksandar Djordjevic [4], Danica Jović [4], Ante Lončarić [1], Jurislav Babić [1], Antun Jozinović [1], Tamara Krska [2], Johann Gangl [5], Chibundu N. Ezekiel [2,6], Michael Sulyok [2] and Rudolf Krska [2,7]

[1] Faculty of Food Technology, Josip Juraj Strossmayer University of Osijek, Franje Kuhača 20, 31000 Osijek, Croatia; bsarkanj@unin.hr (B.Š.); ante.loncaric@ptfos.hr (A.L.); jurislav.babic@ptfos.hr (J.B.); antun.jozinovic@ptfos.hr (A.J.)

[2] Institute of Bioanalytics and Agro-Metabolomics, Department of Agrobiotechnology (IFA-Tulln), University of Natural Resources and Life Sciences Vienna (BOKU), Konrad Lorenzstr. 20, 3430 Tulln, Austria; tamara.krska@gmx.at (T.K.); chaugez@gmail.com (C.N.E.); michael.sulyok@boku.ac.at (M.S.); rudolf.krska@boku.ac.at (R.K.)

[3] Department of Food Technology, University North, Trg dr. Žarka Dolinara 1, 48000 Koprivnica, Croatia

[4] Department of Chemistry, Biochemistry and Environmental Protection, Faculty of Sciences, University of Novi Sad, Trg Dositeja Obradovića 3, 21000 Novi Sad, Serbia; ivana.borisev@dh.uns.ac.rs (I.B.); aleksandar.djordjevic@dh.uns.ac.rs (A.D.); danica.jovic@dh.uns.ac.rs (D.J.)

[5] Institute of Biotechnology in Plant Production, Department of Agrobiotechnology (IFA-Tulln), University of Natural Resources and Life Sciences Vienna (BOKU), Konrad Lorenzstr. 20, 3430 Tulln, Austria; johann.gangl@boku.ac.at

[6] Department of Microbiology, Babcock University, Ilishan Remo 121103, Ogun State, Nigeria

[7] Institute for Global Food Security, School of Biological Sciences, Queen's University Belfast, University Road, Belfast BT7 1NN, Northern Ireland, UK

* Correspondence: tihomir.kovac@ptfos.hr; Tel.: +385-31-224-341; Fax: +385-31-207-115

† These authors contributed equally to this work.

Received: 24 February 2020; Accepted: 25 March 2020; Published: 27 March 2020

Abstract: Despite the efforts to control mycotoxin contamination worldwide, extensive contamination has been reported to occur in food and feed. The contamination is even more intense due to climate changes and different stressors. This study examined the impact of fullerol $C_{60}(OH)_{24}$ nanoparticles (FNP) (at 0, 1, 10, 100, and 1000 ng mL^{-1}) on the secondary metabolite profile of the most relevant foodborne mycotoxigenic fungi from genera *Aspergillus*, *Fusarium*, *Alternaria* and *Penicillium*, during growth in vitro. Fungi were grown in liquid RPMI 1640 media for 72 h at 29 °C, and metabolites were investigated by the LC-MS/MS dilute and shoot multimycotoxin method. Exposure to FNP showed great potential in decreasing the concentrations of 35 secondary metabolites; the decreases were dependent on FNP concentration and fungal genus. These results are a relevant guide for future examination of fungi-FNP interactions in environmental conditions. The aim is to establish the exact mechanism of FNP action and determine the impact such interactions have on food and feed safety.

Keywords: fullerol $C_{60}(OH)_{24}$; nanoparticles; foodborne mycotoxigenic fungi; mycotoxins; secondary metabolism; *Aspergillus* spp.; *Fusarium* spp.; *Alternaria* spp.; *Penicillium* spp.

Key Contribution: The FNP affects the secondary metabolite profile of foodborne mycotoxigenic fungi from the genera *Aspergillus*, *Fusarium*, *Alternaria* and *Penicillium* during growth in vitro. A reduction of concentrations in 35 secondary metabolites was observed; depending on the applied FNP concentration and fungal genus.

---

1. Introduction

The stable water-dissolved forms of fullerene C_{60} (nC_{60}) are nanosized (1–100 nm). During its progression through different environmentally relevant routes, upon photoexcitation with oxygen, nC_{60} form water-soluble oxidised derivatives known as fullerols [1]. Moreover, fullerenes and fullerols are synthesised for specific applications in various industrial commodities [2–6]. Accordingly, the potential for fullerene (nano)materials to enter natural systems is increasing [7–9], which makes fullerol an environmentally relevant daughter product. Despite our efforts, their environmental reactivity, at least in the case of interaction with mycotoxigenic fungi, remains poorly defined [10–13].

It is common knowledge that mycotoxins are not desirable due to their worldwide negative impacts on human and animal health, economies and trade [14]. According to Eskola et al. [15], global mycotoxin prevalence is up to 60%–80%, and such a high occurrence can be explained by a combination of improved sensitivity in analytical methods, and the impact of climate change. The most relevant foodborne mycotoxin producing fungi belong the genera *Aspergillus*, *Fusarium*, *Alternaria* and *Penicillium* [16,17], and maximum levels of secondary metabolites produced by these genera are regulated by the European Union in the Commission Regulation (EC) No. 1881/2006 [18] and the Commission Recommendation 2013/165/EU [19]. Furthermore, plants and fungi can biologically modify mycotoxins by conjugation, while modifications are also possible during food processing. Today these processes contribute, at a significant rate, to food and feed contamination [16]. Accordingly, the European Food Safety Authority is trying to assess exposure and the effects on human and animal health by generating a more accurate database on the occurrence of modified mycotoxins in food and feed [20,21]. The mycotoxigenic fungal community structure has been changing due to climate change, which has put an even bigger emphasis on investigating mycotoxin occurrence. In short, mycotoxin growth and production by major foodborne fungi are highly influenced by climate change factors [22–26], for example, environmental temperature change is affecting contamination by aflatoxins in Eastern Europe, the Balkan Peninsula and Mediterranean regions [27].

Fungal secondary metabolism comprises part of their oxidative stress response pathways. Therefore, stressors to fungi from the environment, including the surrounding microbiome (biotic stressors) or drought, pH, light, neglected environmental compounds, etc. (abiotic stressors), reflect on the reactive oxygen species, and more specifically, the signals (nuclear factors) that initiate/modulate biosynthesis of secondary metabolites, if the stressor is within the level that induces adaptation and survival rather than cell death [10,28,29]. At the same time, fullerol $C_{60}(OH)_{24}$ nanoparticles (FNP) possess an antioxidative potential and poorly defined environmental reactivity [10,12,30]. These factors suggest that FNP modulation of the secondary metabolite profile of mycotoxigenic fungi would likely add to already increased contamination of the environment by mycotoxins.

The aim of this study was, therefore, to determine the effect of FNP on the secondary metabolite profile of the main foodborne mycotoxigenic fungi. It is hypothesised that FNP will modulate secondary metabolism of tested fungi, depending on the applied concentration, while an altered effect dependent on the fungal species is also expected. This study can be accepted as preliminary, as it is expected to be the driving factor for more follow-up-studies on the same issue, with results determining the direction and hypothesis of further research. It is one very important step closer to determine the exact mechanism of FNP action.

2. Results

2.1. Fullerol $C_{60}(OH)_{24}$ Nanoparticle Characterisation

The results of FNP characterisation were obtained through transmission electron microscopy (TEM), dynamic light scattering (DLS) and zeta (ζ) potential measurements (Figure 1). TEM analyses showed (Figure 1a,b) that FNP aqueous solution had particles with sizes less than 10 nm as most dominant, with a tendency to form bigger agglomerates with dimensions under 100 nm. Figure 1c,d represents results of particle size distribution by the number showing that most of the FNP had the mean hydrodynamic radii in the range of 4–18 nm, having a most of the particles (27%) with a size of 7 nm (Figure 1c). The mean ζ potential value of analysed samples was −44.1 mV (Figure 1d).

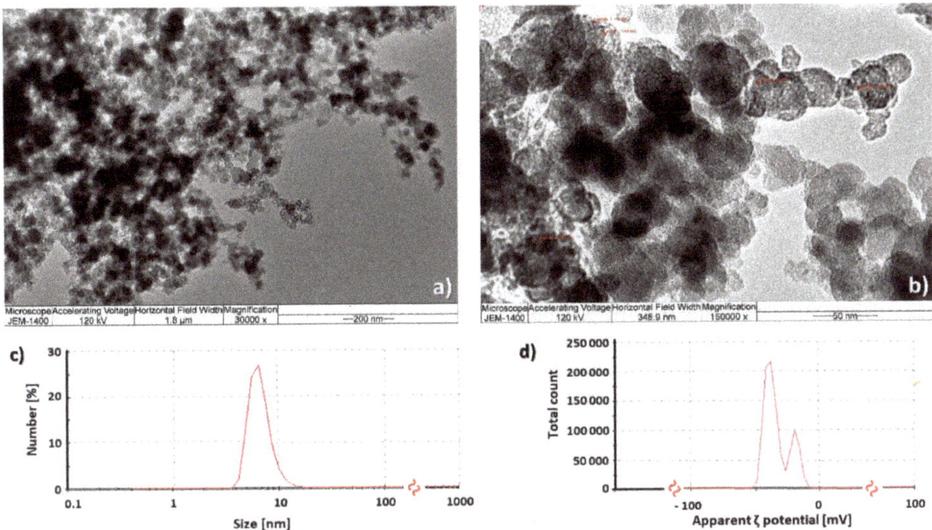

Figure 1. Fullerol $C_{60}(OH)_{24}$ aqueous nanoparticle solution (c = 10 mg mL^{-1}) under transmission electron microscope (TEM) at 30,000× (**a**) and 150,000× (**b**) magnification, as well as particle size distribution by number (**c**) and apparent zeta potential (ζ) (**d**). The data represent one selected result out of three measurements and represent the mean hydrodynamic radius (**c**) and surface charge (**d**). -≈- values on *x*-axes not showed between 100–1000 on (**c**) and 0–100 on (**d**).

2.2. Impact of Fullerol $C_{60}(OH)_{24}$ Nanoparticles on the Growth of Foodborne Mycotoxigenic Fungi

The FNP effect (0, 10, 100 and 1000 ng mL^{-1}) on the growth of tested fungi is presented in Figure 2. After a 72 h growth period at 29 °C in RPMI 1640 media, no statistically significant difference of growth rate between treated and nontreated samples was observed, except in the case of *Fusarium graminearum* mycelia treated with the highest tested FNP dose of 1000 ng mL^{-1} ($p = 0.03$). Furthermore, a statistically significant difference in growth rate ($p = 0.0465$) was observed between FNP applied at 10 and 1000 ng mL^{-1} in the case of *Penicillium expansum*. The dose-dependent growth increase was observed in the case of *Alternaria alternata* (16% to 42%) and *P. expansum* (17% and 32% at FNP of 1 and 10 ng mL^{-1}). The same dose-dependent effect, but of the opposite trend, was observed for *Aspergillus flavus* (7% to 15%), *P. expansum* (6% and 8%, only at applied FNP of 100 and 1000 ng mL^{-1}) and *Fusarium* fungi, at the highest rate. *Fusarium verticillioides* growth rate decreased by 2% to 9% and *Fusarium culmorum* decreased by 12% to 22%, while *F. graminearum* growth rate was decreased at the highest rate observed in this study, from 44% to 50%.

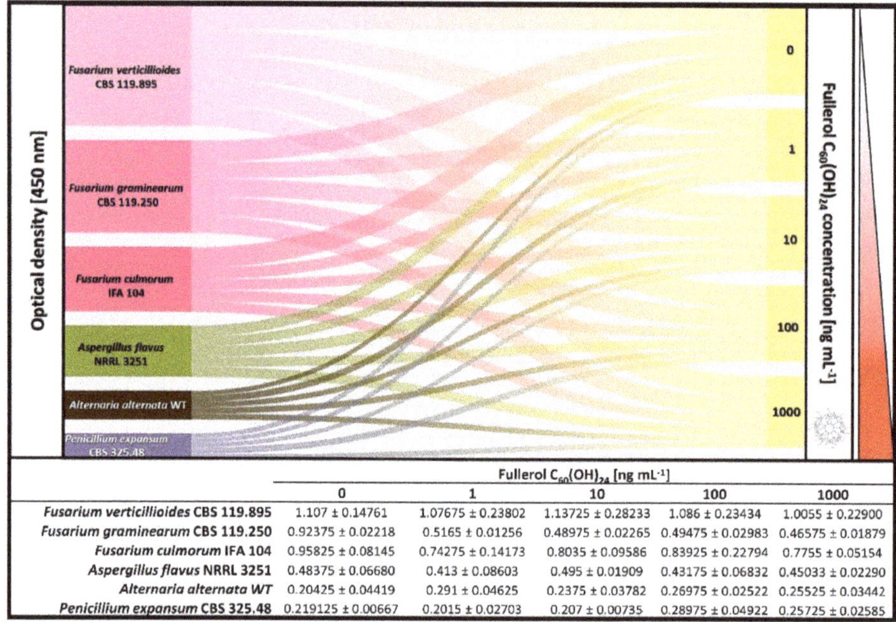

	Fullerol $C_{60}(OH)_{24}$ [ng mL^{-1}]				
	0	**1**	**10**	**100**	**1000**
Fusarium verticillioides **CBS 119.895**	1.107 ± 0.14761	1.07675 ± 0.23802	1.13725 ± 0.28233	1.086 ± 0.23434	1.0055 ± 0.22900
Fusarium graminearum **CBS 119.250**	0.92375 ± 0.02218	0.5165 ± 0.01256	0.48975 ± 0.02265	0.49475 ± 0.02983	0.46575 ± 0.01879
Fusarium culmorum **IFA 104**	0.95825 ± 0.08145	0.74275 ± 0.14173	0.8035 ± 0.09586	0.83925 ± 0.22794	0.7755 ± 0.05154
Aspergillus flavus **NRRL 3251**	0.48375 ± 0.06680	0.413 ± 0.08603	0.495 ± 0.01909	0.43175 ± 0.06832	0.45033 ± 0.02290
Alternaria alternata **WT**	0.20425 ± 0.04419	0.291 ± 0.04625	0.2375 ± 0.03782	0.26975 ± 0.02522	0.25525 ± 0.03442
Penicillium expansum **CBS 325.48**	0.219125 ± 0.00667	0.2015 ± 0.02703	0.207 ± 0.00735	0.28975 ± 0.04922	0.25725 ± 0.02585

Figure 2. Fullerol $C_{60}(OH)_{24}$ nanoparticles (FNP) influence on ▪ *Fusarium verticillioides* (CBS 119.825), ▪ *Fusarium graminearum* (CBS 110.250), ▪ *Fusarium culmorum* (IFA 104), ▪ *Aspergillus flavus* (NRRL 3251), ▪ *Alternaria alternata* (WT) and ▪ *Penicillium expansum* (CBS 325.48) biomass production (expressed as optical density at 450 nm) in liquid RPMI 1640 medium incubated over a 72 h period at 29 °C. Data represent the mean ± standard error of the mean (SEM) from three separate experiments.

2.3. The Impact of Fullerol $C_{60}(OH)_{24}$ Nanoparticles on Secondary Metabolite Profiles of Selected Foodborne Mycotoxigenic Fungi

The FNP effect on the secondary metabolite profiles of selected foodborne mycotoxigenic fungi is presented in Figures 3–6.

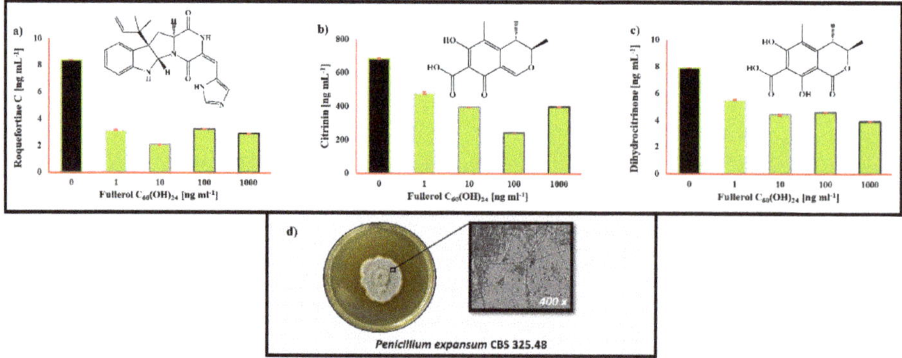

Figure 3. *Penicillium expansum* (CBS 325.48) secondary metabolite profile in liquid RPMI 1640 medium after 72 h at 29 °C influenced by 0, 1, 10, 100 and 1000 ng mL^{-1} of fullerol $C_{60}(OH)_{24}$ nanoparticles (FNP). Detected were (**a**) roquefortine C, (**b**) citrinin and (**c**) dihydrocitrinone. Data represent the mean ± standard error of the mean (SEM) from three separate experiments and are expressed in ng mL^{-1}. *P. expansum* colony was grown on potato dextrose agar for 168 h (**d**).

In RPMI 1640 media, *P. expansum* (Figure 3) produced the following mycotoxins: roquefortine C, citrinin and dihydrocitrinone. These compounds were observed in our control and FNP-treated samples, however, a decrease in the production of these compounds was observed after FNP treatments. For example, roquefortine C exhibited a greater decrease in concentration (61% to 75%). The concentration of produced citrinin exhibited an FNP-related decrease of 30% to 65%, caused by FNP at 100 ng mL^{-1}. The dihydrocitrinone concentration decreased by 30% to 51%. There were also observed statistically significant ($p = 0.01$) differences between the control and FNP-treated samples, including the roquefortine C concentration at 10 ng mL^{-1} of FNP, the citrinin concentration at 100 ng mL^{-1} of FNP, as well as the dihydrocitrinone concentration at 1000 ng mL^{-1} of FNP.

Alternaria alternata secondary metabolites produced in the RPMI 1640 media under the presence of FNP are shown in Figure 4. The presence of alternariol, alternariolmethylether and tenuazonic acid were observed, both in control and FNP treated samples. Again, in the case of all mentioned metabolites, a decrease of produced concentrations was observed, after treatment of fungi with FNP during growth, in comparison with control samples. Alternariol exhibited a concentration decrease of 11%–65% upon exposure to FNP, whereas reductions for alternariolmethylether and tenuazonic acid were reduced up to 100% and 66%, respectively. Statistically significant differences between the control and FNP-treated samples were also observed, including alternariol at 10 ng mL^{-1} of FNP ($p = 0.026$), alternariolmethylether at 1000 ng mL^{-1} ($p = 0.01$) and tenuazonic acid at 1 ng mL^{-1} ($p = 0.01$).

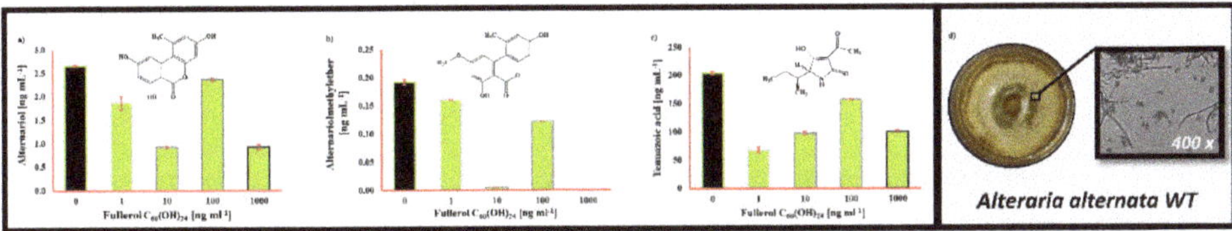

Figure 4. *Alternaria alternata* (WT) secondary metabolite profile in liquid RPMI 1640 medium after 72 h at 29 °C influenced by 0, 1, 10, 100 and 1000 ng mL^{-1} of fullerol C$_{60}$(OH)$_{24}$ nanoparticles (FNP). Detected were (**a**) alternariol, (**b**) alternariolmethylether and (**c**) tenuazonic acid. Data represent the mean ± standard error of the mean (SEM) from three separate experiments and are expressed in ng mL^{-1}. *A. alternata* (WT) colony was grown on potato dextrose agar for 168 h (**d**).

For the *Aspergillus flavus* secondary metabolites produced under the presence of FNP, in the RPMI 1640 media, aflatoxin B1 (AFB1), kojic acid, norsorolinic acid, cyclopiazonic acid, 3-nitropropionic acid, asperfuran and dichlordiaportin were observed both in the control and FNP treated samples (Figure 5). While AFB1, kojic acid, cyclopiazonic acid, 3-nitropropionic acid and dichlordiaportin exhibited an overall decrease in concentrations if FNP was applied, norsorolinic acid and asperfuran generally exhibited an increase in concentration during FNP exposure.

FNP decreased AFB$_1$ concentration 59%–75% and decreased 3-nitropropionic acid production by up to 58% (Figure 5). The respective reductions observed for dichlordiaportin, kojic acid and cyclopiazonic acid were found to range from 16%–24%, 50%–72% and 24%–70%. The two *A. flavus* mycotoxins that exhibited highly variable FNP-related responses (decreases and increases in concentrations) were norsolorinic acid and asperfuran. Norsolorinic acid concentrations increased by 17% and 54% (at 1 and 10 ng mL^{-1} of FNP, respectively), while 100 and 1000 ng mL^{-1} of FNP decreased its concentrations by 28% and 7%, respectively. Respective asperfuran concentrations decreased by 22% and 20% (at 1 and 1000 ng mL^{-1}), but they also increased by 43% and 33% at respective FNP concentrations of 10 and 100 ng mL^{-1}. AFB$_1$ concentration was significantly decreased at 1 ng mL^{-1} ($p = 0.010$), dichorodiaportin at 100 ng mL^{-1} ($p = 0.014$), kojic acid at 1 ng mL^{-1} ($p = 0.010$) while cyclopiazonic acid concentration was statistically significant decreased at 1000 ng mL^{-1} ($p = 0.010$) of FNP. Moreover, a statistically significant difference ($p = 0.010$) in FNP effect was observed at norsorolinic acid between

10 and 100 ng mL^{-1} of FNP. Asperfuran concentrations were statistically significant different after application of 1 and 10 ng mL^{-1} of FNP ($p = 0.035$) and 1 and 1000 ng mL^{-1} of FNP ($p = 0.047$).

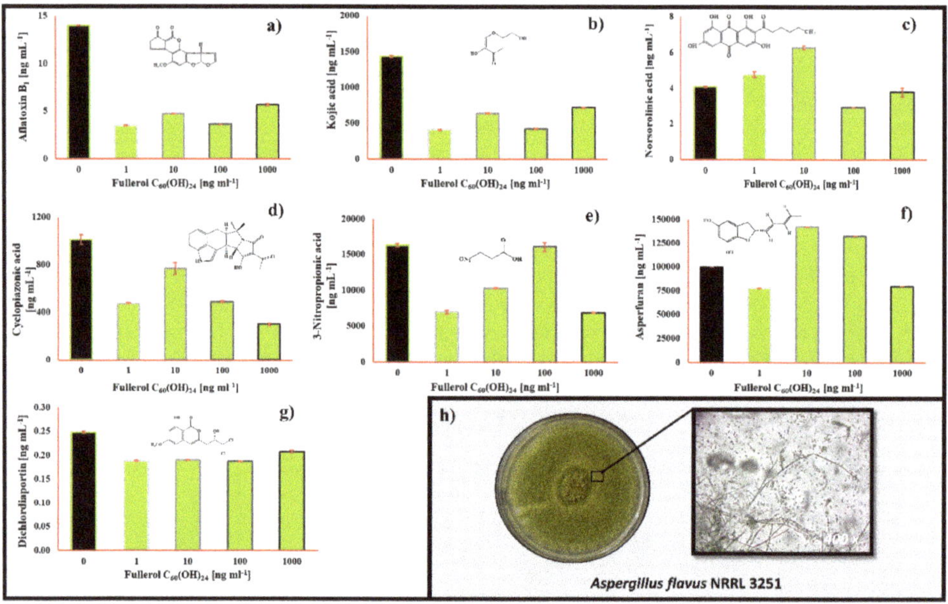

Figure 5. *Aspergillus flavus* (NRRL 3251) secondary metabolites profile in liquid RPMI 1640 medium after 72 h at 29 °C influenced by 0, 1, 10, 100 and 1000 ng mL^{-1} of fullerol C$_{60}$(OH)$_{24}$ nanoparticles (FNP) (**a–g**). Data represent the mean ± standard error of the mean (SEM) from three separate experiments and are expressed in ng mL^{-1}. *A. flavus* NRRL 3251 colony was grown on potato dextrose agar for 168 h (**h**).

In Figure 6, the secondary metabolite profile of selected *Fusarium* spp. fungi grown in RPMI 1640 medium during 72 h at 29 °C influenced by 0, 1, 10, 100 and 1000 ng mL^{-1} of FNP is presented. Figure 6a presents the *F. verticillioides* secondary metabolites produced under the presence of FNP. In the RPMI 1640 media, the presence of aurofusarin, culmorin, fusaric acid, sambucinol, equisetin, fusarin C and gibepyron D were observed, both in the control and FNP treated samples, except equisetin. Interestingly, no major trichothecenes were detected in the RPMI 1640 medium (deoxynivalenol, nivalenol, T-2 toxin, HT-2 toxin and their derivatives), together with fumonisins. FNP did not affect aurofusarin at 1 ng mL^{-1}, but they completely blocked its production, or secretion in media, at all other tested concentrations. When culmorin was present, there was a certain increase in production, from 7% to 67%, whereas the highest increase was in the presence of FNP at 1 ng mL^{-1}, again. Similarly, the fusaric acid concentrations increased even more, from 128% to 542%, at the 1 ng mL^{-1} treatment with FNP. In the case of sambucinol, the 1 and 100 ng mL^{-1} treatments caused respective increases in concentration of 94% and 133%, while 10 and 1000 ng mL^{-1} caused respective concentration decreases of 50% and 11%. Equisetin was not detected in the control samples when FNP at 1 and 10 ng mL^{-1} was applied, but it was observed after addition of 100 and 1000 ng mL^{-1} of FNP in growth media. Fusarin C concentrations were decreased by FNP, from 8% to 22%, while 10 ng mL^{-1} of FNP caused a 22% decrease. The FNP caused a decrease of gibepyron D concentrations from 58% to 64%, while its concentration was also increased 8% after addition of 10 ng mL^{-1} in growth media. Furthermore, the statistically significant differences were observed between control and FNP treated samples but also between applied FNP concentrations. Differences in culmorin and fusaric acid production were statistically significant, in comparison with control samples, after the addition of 1 ng mL^{-1} of FNP ($p = 0.01$). The fusarin C concentration was statistically significant in comparison with control samples when 10 ng mL^{-1} of FNP applied ($p = 0.01$). Moreover, there was a significant difference between the effect of FNP at 10 and 100 ng mL^{-1} on sambucinol ($p = 0.01$), while in the case of gibepyron D, a significant difference was noted between 1 and 10 ng mL^{-1}.

Fusarium graminearum secondary metabolites produced in the RPMI 1640 media under the presence of FNP (Figure 6b) were: aurofusarin, chrysogin, culmorin, rubrofusarin, sambucinol, butenolid and zearalenone-sulfate, both in the control and FNP treated samples. FNP present in growth media at all tested concentrations reduced aurofusarin concentrations by 40%–43%, while chrysogin was reduced by 61%–65%. Culmorin concentrations were decreased by 8% to 55%, while the most effective inhibition of 55% was noted when 100 ng mL^{-1} of FNP was applied. At the same time, when the 1000 ng mL^{-1} of FNP was applied, culmorin concentration in growth media was decreased by 24%. Similar to that, rubrofusarin concentration was increased by 34% after the application of 1000 ng mL^{-1} of FNP, while at all other tested FNP doses a decrease in concentration was noted, from 26%–51%; the highest decrease when 10 ng mL^{-1} of FNP was present in the growth media. Alternatively, the highest decrease in butenolid concentrations was noted at 10 ng mL^{-1} of FNP (33%), while the decrease ranged from 27%–33%, in general. The zearalenone-sulfate concentrations were decreased by 41%–55%, at the highest rate of 55% when 100 ng mL^{-1} of FNP was present in growth media. The sambucinol concentration was decreased at all tested FNP concentrations, from 145%–984%, while 10 ng mL^{-1} of FNP caused the highest decrease of 984%. As mentioned above, not only were there statistically significant differences between control and FNP treated samples, but also between applied FNP concentrations. The aurofusarin concentration after addition of 10 ng mL^{-1} of FNP in growth media was statistically significant different ($p = 0.01$) in comparison with the control sample. The sambucinol and butenolid production was statistically significant different, after addition of 10 ng mL^{-1} of FNP ($p = 0.01$), while zearalenone-sulfate concentrations were significantly different at 100 ng mL^{-1} of FNP ($p = 0.01$). The chrysogin concentration produced under the effect of 1000 ng mL^{-1} was also statistically significantly different in comparison with control samples ($p = 0.026$). Statistically significant difference between the applied FNP concentrations was noted at culmorin between 100 and 1000 ng mL^{-1} of FNP ($p = 0.01$), while at produced rubrofusarin concentrations difference was noted between 10 and 1000 ng mL^{-1} of FNP ($p = 0.01$).

Figure 6c shows the *Fusarium culmorum* secondary metabolites produced under the presence of FNP. In the RPMI 1640 media, the presence of aurofusarin, beauvericin, chrysogin, culmorin, sambucinol, zearalenone and zearalenone- sulfate were detected, both in control and FNP treated samples. FNP reduced aurofusarin by 44% to 58% at all tested FNP concentrations, except 100 ng mL^{-1}, which increased its concentration in growth media by 52%. Chrysogin concentration was reduced by FNP from 50%–54%, zearalenone from 79% to 86% and zearalenone-sulfate from 48%–55%. On the other hand, in the case of beauvericin, FNP increased production by 228% to 1342%, while the highest increase was caused by an FNP concentration of 100 ng mL^{-1}. Moreover, culmorin concentrations were also increased in FNP presence, from 36%–63%, where FNP at 1 ng mL^{-1} caused the highest increase of 63%. As mentioned above, between *F. culmorum* secondary metabolites were again observed statistically significant differences between control and FNP treated samples, as well as between applied FNP concentrations. The beauvericin concentration after addition of 100 ng mL^{-1} of FNP in growth media was significantly ($p = 0.01$) different in comparison with the control sample, while in the chrysogin and zearalenone-sulfate concentrations differences ($p = 0.01$) were observed in the case of 10 ng mL^{-1} of FNP, culmorin at 1 ng mL^{-1} of FNP ($p = 0.01$) and zearalenone at 1000 ng mL^{-1} of FNP ($p = 0.01$). Moreover, a statistically significant difference between the applied FNP concentrations was noted in the case of aurofusarin and sambucinol at 1 and 100 ng mL^{-1} of FNP present in growth media ($p = 0.01$).

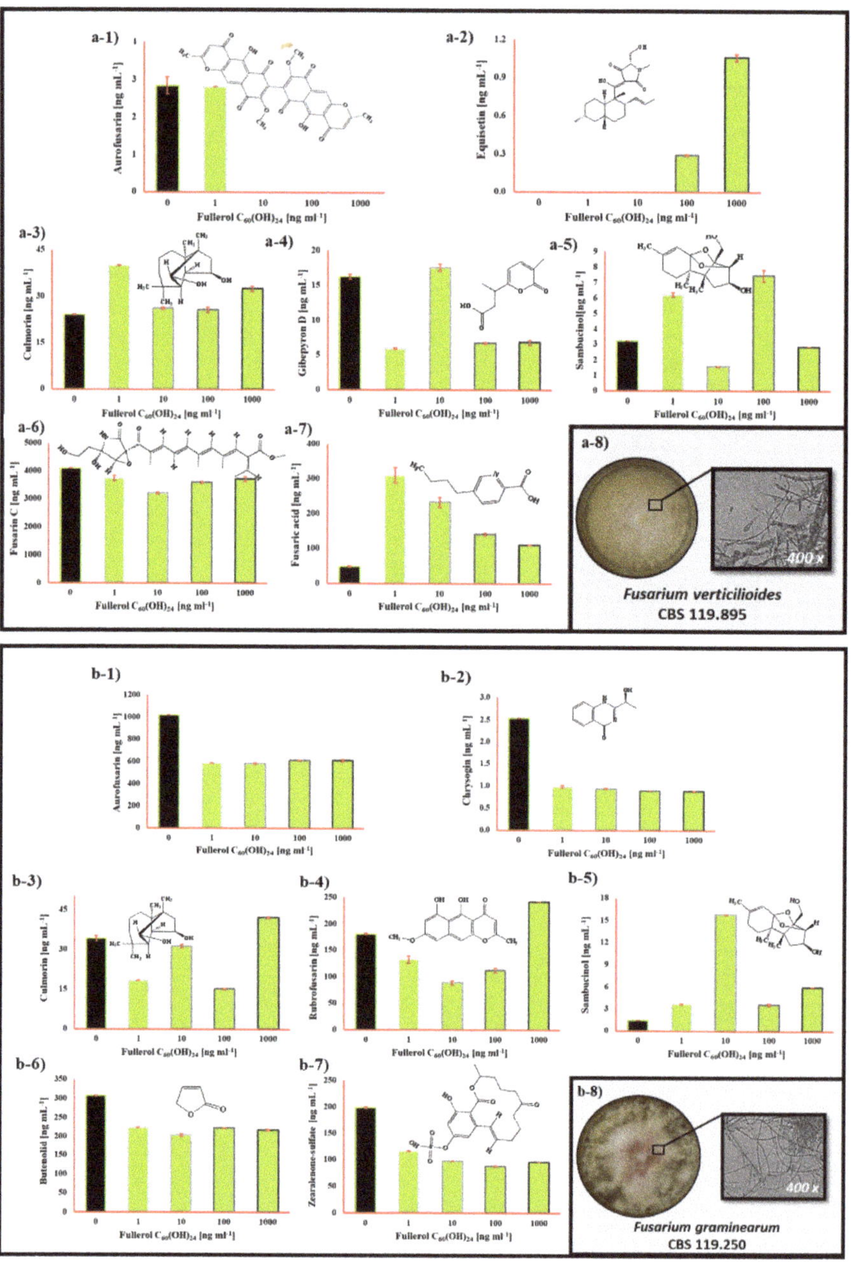

Figure 6. *Cont.*

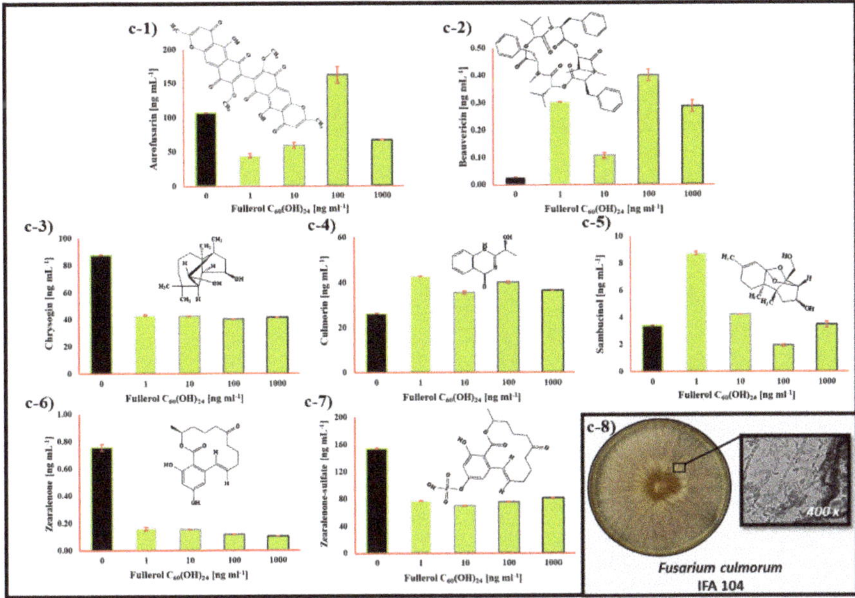

Figure 6. (**a**) *Fusarium verticillioides* (CBS119.825), (**b**) *Fusarium graminearum* (CBS 110.250) and (**c**) *Fusarium culmorum* (IFA 104) secondary metabolites profile in liquid RPMI 1640 medium after 72 h at 29 °C influenced by 0, 1, 10, 100 and 1000 ng mL^{-1} of fullerol $C_{60}(OH)_{24}$ nanoparticles (FNP). The detected *F. verticillioides* (CBS119.825) metabolite were (**a-1**) aurofusarin, (**a-2**) equisetin, (**a-3**) culmorin, (**a-4**) gibepyron D, (**a-5**) sambucinol, (**a-6**) fusarin C and (**a-7**) fusaric acid. *F. verticillioides* colony grown on potato dextrose agar for 168 h (**a-8**). The detected *Fusarium graminearum* (CBS 110.250) metabolites were (**b-1**) aurofusarin, (**b-2**) chrysogin, (**b-3**) culmorin, (**a-4**) rubrofusarin, (**b-5**) sambucinol, (**b-6**) butenolid and (**b-7**) zearalenone-sulfate. *F. graminearum* colony grown on potato dextrose agar for 168 h (**b-8**). The detected *F. culmorum* (IFA 104) metabolite were (**c-1**) aurofusarin, (**c-2**) beauvericin, (**c-3**) chrysogin, (**c-4**) culmorin, (**c-5**) sambucinol, (**c-6**) zearalenone and (**c-7**) zearalenone-sulfate. *F. culmorum* colony was grown on potato dextrose agar for 168 h (**c-8**). Data of all tested *Fusarium* fungi represent the mean ± standard error of the mean (SEM) from three separate experiments and are expressed in ng mL^{-1}.

3. Discussion

On the basis of the previously reported FNP effect on *A. flavus* [10,12], our goal was to determine if there is a similar effect of FNP on the secondary metabolism of other important foodborne mycotoxigenic fungi. The hypothesis was that FNP would modulate secondary metabolism of tested fungi, depending on the applied concentration, while the effect will also be genus- or species-dependent. The results herein establish the foundation for follow-up studies to determine the FNP mechanism during interaction with mycotoxigenic fungi.

The size of nanoparticles, as well as the ability to form aggregates/agglomerates, are the features that determine their properties and subsequent biological activity. Thus, results of TEM, DLS and ζ potential indicate that the prepared FNP water solution is a stable polyanionic system mostly composed of clusters with sizes less than 100 nm. These results are in accordance with our previously published results and EC Recommendation for the definition of the term "nanomaterial" [10,12,31–34].

In general, our findings are in accordance with previously reported data on FNP impact on mycotoxigenic fungi biomass production. However, available data stems from experiments only conducted with *A. flavus* [10,12], and to the best of our knowledge, this is the first report on FNP impact on representatives of *Alternaria*, *Penicillium* and *Fusarium* spp. It is evident that every tested concentration of FNP yielded no strong antifungal effect. Despite the above-mentioned differences, there were certain effects of FNP on the foodborne mycotoxigenic fungi growth rate, which is in accordance with a previous study conducted by Holmes et al. [35]. Our findings suggest that his conclusions can be applied to other mycotoxigenic fungal species. Indeed, the confirmed FNP dose-dependent effect on *A. flavus* biomass in different liquid growth media extends to other tested fungal species [10,12]. Some of the inhibitors at lower concentrations can modulate secondary metabolism, but at higher concentrations, they can affect the growth of fungi. Moreover, growth inhibition cannot be directly correlated with secondary metabolism intensity, but it is important to examine both parameters. For example, the fact that growth of fungi under abiotic stressors remains practically unaffected is in accordance with results of Medina et al. [26,36] and Kovač et al. [12]. It is known that many compounds with inhibitory potential are also able to act as antioxidants, although the mode of their action is poorly understood [35]. We have shown this to also be true for FNP, which exhibit antioxidative properties. Furthermore, it is known fact that mycotoxigenic fungi are very sensitive to oxidative status perturbations, and one of their defence mechanisms involves production and export of secondary metabolites out of the cell, at least in the case of *A. flavus* [10,29].

Penicillium expansum produced secondary metabolites under the presence of FNP, as shown in Figure 3. Roquefortine C (Figure 3a) is a modified diketopiperazine, which showed bacteriostatic activity against G+ bacteria and also can interact with cytochrome p450 and interfere with RNA synthesis; this mycotoxin is also neurotoxic to cockerels [37]. Citrinin (Figure 3b) is a polyketide compound involved in the etymology of Balkan endemic nephropathy. Toxicity and genotoxicity of citrinin are related to oxidative stress or increased permeability of mitochondrial membranes. In mice and rabbits, LD_{50} is estimated at 100 mg kg^{-1} BW, while citrinin residues may occur in edible tissues and eggs after oral exposure of animals to highly contaminated feed [38,39]. Dihydrocitrinone (Figure 3c) is a secondary metabolite that originated from *Penicillium* and *Aspergillus* spp. and is related to citrinin. To be more precise, it is the less toxic metabolite of citrinin which is found in the urine and blood of rats and humans [40]. All the mentioned metabolites of *P. expansum* were detected after growth in RPMI 1640 media, with roquefortine C, citrinin and dihydrocitrinone, exhibiting a decrease in concentration after FNP was added to the growth media.

Alternaria alternata secondary metabolites produced under the presence of FNP are shown in Figure 4. Alternariol, alternariolmethylether and tenuazonic acid were all detected in the growth media. Alternariol (Figure 4a) and alternariolmethylether (Figure 4b) are dibenzo-α-pyrones, and it is believed that they are mutagenic because of their in vitro genotoxic effects. It has been shown that they interfere with topoisomerases I and II, with effects that have been described as 'poisoning'. Tenuazonic acid (Figure 4c) belongs to a group of tetramic acid derivatives and is acutely toxic (LD_{50} 81–225 mg kg^{-1} BW; mice). It is produced as a virulence factor to facilitate the colonisation of the fungus on plants since it inhibits protein biosynthesis by suppressing the release of new proteins from the ribosomes [41]. All of the mentioned *A. alternata* metabolites detected after growth in RPMI 1640 media were decreased in concentration after FNP was added to the growth media.

Aspergillus flavus secondary metabolites were produced under the presence of FNP in RPMI 1640 media, as shown in Figure 5. AFB_1, kojic acid, norsorolinic acid, cyclopiazonic acid, 3-nitropropionic acid, asperfuran and dichlordiaportin were detected in the growth media. The impact of FNP on the *A. flavus* secondary metabolite profile was already reported [10,12]. However, these experiments were conducted in different microbiological media, so these results are the first generated from *A. flavus* grown in RPMI 1640 media for this purpose, but are in still in accordance with previous reports. Again, there was observed reduction of secondary metabolite concentrations in growth media as we already stated [10,12,13], which is proof of a reasonably conducted and well-designed study. Accordingly, the detected metabolites will not be explained in detail because they can be found in recently published studies.

Secondary metabolites produced by selected *Fusarium* spp. fungi are shown in Figure 6. The results confirm that most of the secondary metabolite pathways are specific to the species [42]. All three species produced aurofusarin, culmorin and sambucinol (Figure 6a–c), equisetin, gibepyron D, fusarin C, while fusaric acid was produced only by *F. verticillioides* (Figure 6a). At the same time, chrysogin, rubrofusarin, butenolid and zearalenone-sulfate were produced by *F. graminearum* (Figure 6b), while beauvericin and zearalenone were produced by *F. culmorum* (Figure 6c). The aurofusarin is dimeric naphtho-γ-pyrone and is produced by many ascomycetes where predation and mechanical damage stimulate aurofusarin synthesis. It is a red pigment known from pure culture of the fungi (Figure 6b-8)) or from maize ears that are infected with *F. graminearum* [43–45]. In general, aurofusarin concentration in growth media was decreased at all isolates, except when 100 ng mL^{-1} of FNP was applied during *F. verticillioides* growth (Figure 6c-1). Furthermore, sambucinol and culmorin are increased in concentration, except in the case of culmorin produced by *F. graminearum* (Figure 6). Culmorin is an 'emerging toxin' with a sesquiterpene diol core structure, not so much investigated and produced by *Fusarium* spp. fungi, *F. culmorum* (name giving) for example. Its toxicological properties are poorly described in the literature, but the LD$_{50}$ is estimated for mice from 250–1000 µg kg^{-1} BW [46]. At the moment, the role of sambucinol, a major metabolite of *F. culmorum*, is unknown. However, there is evidence that culmorin can seriously affect the deoxynivalenol metabolism in mammals by suppressing glucuronidation [47]. Equisetin is a secondary metabolite of *F. verticillioides* detected in growth media after addition of 100 and 1000 ng mL^{-1} of FNP (Figure 6a-2). It is a member of the acyl tetramic acid family of natural products. There are reports about its biological activity as an antibiotic and for its HIV inhibitory activity, cytotoxicity and mammalian DNA binding. For those reasons, it is synthesised under laboratory conditions [48]. Furthermore, the observed concentrations of gibepyron D, known for its nematocidal activity [49], were dependent upon added FNP dose (Figure 6a-4). Fusarin C, a metabolite that shows mycoestrogenic properties [50], is slightly decreased or unaffected in concentration after addition of FNP (Figure 6a-6). Fusaric acid (Figure 6a-7), picolinic acid derivative produced by *Fusarium* spp. fungi, responsible for induced programmed cell death and altered MAPK signalling in healthy PBMCs and Thp-1 cells [51], showed an increase in concertation after all tested FNP doses. The chrysogin, butenolid, rubrofusarin, zearalenone, beauvercin and zearalenone-sulfate, *F. graminearum* and *F. culmorum* metabolites, were also detected in growth media, as already mentioned. While there are no reports about chrysogin activity in the scientific literature, butenolid is reported as a myocardial mitochondria dysfunction inducer, while rubrofusarin is a pigment and an intermediate in the aurofusarin biosynthesis pathway [52]. Zearalenone-sulfate is a modified form of zearalenone, which exhibits low acute toxicity, but acts as a full and partial agonist on oestrogen receptors α and β, and cause reproduction disorders in pigs. Moreover, it is an activator of receptors involved in the regulation of cytochrome P450 isoforms in vitro. Due to that, it potentially affects the phase I metabolism of various endo- and xenobiotics [53]. Beauvericin is a nonribosomal cyclic hexadepsipeptide that possesses insecticidal properties and which can induce apoptosis of mammalian cells [54]. Beauvericin and rubrofusarin concentrations were increased by FNP, while chrysogin, butenolid, zearalenone and zearalenone-sulfate concentrations were decreased in growth media due to FNP presence (Figure 6b,c).

4. Conclusions

The exposure of FNP directly affected mycotoxin concentrations in important foodborne fungi, indicating that FNP definitely modulates fungal secondary metabolism. Our findings support the continued study of FNP impact on mycotoxigenic fungal species. Future studies will explore gene expression, as well as the oxidative status of fungal cells. This will allow us to confirm that the FNP effect on fungi is concentration-dependent and that due to oxidative status, modulation of the cells changes the secondary metabolite profile. This will be helpful in establishing a scientific outline on the nature of the FNP-fungi interaction, and the exact impact FNP will have on food safety.

5. Materials and Methods

5.1. Chemicals

Acetonitrile and methanol (both HPLC grade) were obtained from Merck (Darmstadt, Germany). Ammonium acetate, glacial acetic acid (p.a.), MOPS, and RPMI 1640 medium were purchased from Sigma–Aldrich (Vienna, Austria). Sodium hydroxide was purchased from Panreac (Barcelona, Spain). For ultrapure water preparation, a Purelab ultra system (ELGA LabWater, Celle, Germany) was used. Standards used in the study were prepared according to Malachova et al. [55] and were collected from various research groups, or purchased from the following commercial sources: Romer Labs® Inc. (Tulln, Austria), Sigma–Aldrich (Vienna, Austria), Iris Biotech GmbH (Marktredwitz, Germany), Axxora Europe (Lausanne, Switzerland) and LGC Promochem GmbH (Wesel, Germany).

5.2. Fullerol $C_{60}(OH)_{24}$ Synthesis, Preparation and Characterisation of Nanoparticle Solution

Reagents used in the process of synthesis of fullerol $C_{60}(OH)_{24}$ nanoparticles (FNP) were all analytical grade (C_{60} (99.8% purity), Br_2, NaOH and C_2H_5OH). FNP were synthesised in a two-step synthetic protocol. First, polybromine derivative was synthesised by the reaction of polybromination of C_{60} by Br_2 using $FeCl_3$ as a catalyst, with the aim to obtain polybrominated derivative $C_{60}Br_{24}$ as described in the paper by Djordjevic et al. [56]. The second step included the complete substitution of bromine atoms with OH groups which were achieved in alkaline media by the use of NaOH solution. This procedure was followed by repetitive rinsing of the obtained mixture with ethanol, removing the residual components, and elution with demineralised water, according to the procedure given by Mirkov et al. [57]. The final solution of FNP in water was dried under low pressure until the dark brown powder of the desired substance was obtained. For the purpose of further experiments the calculated amount of FNP powder was dissolved in demineralised water (pH = 6.5), sonicated for 15 min, after which the final solution at a concentration of 10 ppm was obtained.

In order to obtain more complete insight concerning FNP distribution, and to complement the results of DLS measurements, we performed transmission electron microscopy analyses of aqueous FNP solution. The measurements were conducted on a JEM 1400 microscope operating at an accelerating voltage 120 kV, using a horizontal field width of 173.9 nm and magnification of 300,000×. A few drops of FNP water solution sample were applied to copper grid 300 mesh, then dried at standard room temperature and measured.

DLS was used for determination of the hydrodynamic mean diameter of the particles, while zeta potential measurements were conducted for particles surface charge determination. All measurements were performed on a Zetasizer Nano ZS (Malvern Instruments Inc., Malvern, UK), at 633 nm wavelength and 173° measurement angle (backscatter detection) in water solution at room temperature. DLS measurements were conducted in triplicate, and zeta potential measurements were performed in duplicate.

5.3. Tested Fungal Strains, Inoculum Preparation and Cultivation

Fungi used in this study are major producers of mycotoxins and food contaminants [12,58]: *Fusarium verticillioides* (CBS119.825), *Fusarium graminearum* (CBS 110.250), *Fusarium culmorum* (IFA 104), *Aspergillus flavus* (NRRL 3251), *Alternaria alternata* (WT) and *Penicillium expansum* (CBS 325.48).

Inoculum preparation was performed, according to Jerković et al. [59]. The number of conidia in the stock suspension was adjusted on 10^6 CFU mL^{-1} by using a Bürker-Türk counting chamber (Haemocytometer), and checked by inoculating the dilutions to PDA agar. Working $2 \times$ conidia suspension was prepared by diluting 200 µL of stock suspension into 10 mL of RPMI 1640 medium.

Tested fungi were cultivated in the RPMI 1640 medium, buffered with 0.164 M MOPS (34.53 g L^{-1}) and adjusted to pH 7.0 with (1 M) NaOH, as recommended by the National Committee for Clinical Laboratory Standards. The medium was sterilised by filtration using a 0.22 µm bottle top filter. Sterile clear polystyrene microtiter, plates (96 U-shaped wells; Ratiolab, Dreieich, Germany) were used in the microdilution test and inoculated media was amended with sterile FNP stock solution (10 mg mL^{-1}) to a final concentration of 0, 1, 10, 100 and 1000 ng mL^{-1} of FNP solution.

After inoculation plates were incubated at 29 °C in an atmospheric incubator in darkness. The 72 h incubation period was needed to read plates on a microplate reader at 450 nm (Azure boisystems, Ao absorbance microplate reader, Dublin, CA, USA). Minimal inhibitory concentration for 50% cell death (MIC50) was defined as the lowest concentration reducing the optical density by 50% at 450 nm compared with growth control.

5.4. Determination of Fungal Secondary Metabolites in Culture Media

The metabolites produced by tested fungi in culture media were determined by the multi-analyte 'dilute and shoot' LC-MS/MS method of Malachova et al. [55]. For the determination of metabolites in growth media, a ten-fold dilution of 100 µL of the medium with a mixture of extraction solvent and dilution solvent (1:1, v/v) in glass vials without any pre-treatment was performed.

The screening and detection of metabolites were performed, as described by Malachova et al. [55]. At least two sSRM transitions were monitored per metabolite (quantifier and qualifier), and according to the validation guidelines, the ratio between two transitions was used as an additional identity confirmation point. Analyst 1.7.1. was used for qualitative conformation, while Multiquant 3.0.3 (SCIEX, Vienna, Austria). was used for quantification of detected analytes.

5.5. Statistical Analysis

All data were expressed as the mean value ± standard error of the mean (SEM) from three separate experiments. The pooled datasets were checked for normality distribution by Shapiro-Wilk test and compared by nonparametric statistics methods (Friedman ANOVA and Kendall coefficient of concordance; Kruskal-Wallis ANOVA). The programme package Statistica 13.1 (TIBCO Software Inc., Palo Alto, CA, USA) was used, and differences were considered significant when the p-value was <0.05. For the drawing of the Sankey diagrams Flourish studio was used (Flourish Studio, Kiln Enterprises Ltd., London, UK).

Author Contributions: Conceptualization, T.K. (Tihomir Kovač), B.Š., M.S. and R.K.; methodology, T.K. (Tihomir Kovač), B.Š., I.B., D.J., A.D., C.N.E.; software, M.S., T.K. (Tihomir Kovač), B.Š., I.B., D.J., A.D.; formal analysis, T.K. (Tihomir Kovač), B.Š., I.B., D.J., J.G., M.S.; investigation, T.K. (Tihomir Kovač), B.Š.; resources, T.K. (Tihomir Kovač), A.J., A.L., J.B., R.K.; data curation, T.K. (Tihomir Kovač), B.Š.; writing—original draft preparation, T.K. (Tihomir Kovač); writing—review and editing, T.K. (Tihomir Kovač), B.Š., A.L., T.K. (Tamara Krska), C.N.E.; I.B., A.D., M.S., A.J., J.B, R.K.; visualization, T.K. (Tihomir Kovač), B.Š., T.K. (Tamara Krska); supervision, J.B., M.S., R.K. All authors have read and agreed to the published version of the manuscript.

Funding: The synthesis and characterization of fullerol C$_{60}$(OH)$_{24}$ nanoparticles used in this study was supported by a grant from the Ministry of Education, Science and Technological Development of the Republic of Serbia (Grant No. 451-03-68/2020-14/200125).

Acknowledgments: The authors are grateful to Vladimir Pavlović (Faculty of Agriculture, University of Belgrade) for TEM measurements.

References

1. Wu, J.; Alemany, L.B.; Li, W.; Benoit, D.; Fortner, J.D. Photoenhanced transformation of hydroxylated fullerene (fullerol) by free chlorine in water. *Environ. Sci. Nano* **2017**, *4*, 470–479. [CrossRef]
2. Michalitsch, R.; Kallinger, C.; Verbandt, Y.; Veefkind, V.; Huebner, S.R. The fullerene patent landscape in Europe. *Nanotechnol. Law Bus.* **2008**, *5*, 85–94.
3. Duncan, T.V. Applications of nanotechnology in food packaging and food safety: Barrier materials, antimicrobials and sensors. *J. Colloid Interface Sci.* **2011**, *363*, 1–24. [CrossRef]

4. Pycke, B.F.G.; Chao, T.-C.; Herckes, P.; Westerhoff, P.; Halden, R.U. Beyond nC_{60}: Strategies for identification of transformation products of fullerene oxidation in aquatic and biological samples. *Anal. Bioanal. Chem.* **2012**, *404*, 2583–2595. [CrossRef] [PubMed]

5. Dai, L. Synthesis of fullerene- and fullerol-containing polymers. *J. Mater. Chem.* **1998**, *8*, 325–330. [CrossRef]

6. Zhou, A.H.; Zhang, J.D.; Xie, Q.J.; Yao, S.Z. Application of double-impedance system and cyclic voltammetry to study the adsorption of fullerols ($C_{60}(OH)_n$) on biological peptide-adsorbed gold electrode. *Biomaterials* **2001**, *22*, 2515–2524. [CrossRef]

7. Farré, M.; Sanchís, J.; Barceló, D. Analysis and assessment of the occurrence, the fate and the behavior of nanomaterials in the environment. *TrAC Trends Anal. Chem.* **2011**, *30*, 517–527. [CrossRef]

8. Sanchís, J.; Bosch-Orea, C.; Farré, M.; Barceló, D. Nanoparticle tracking analysis characterisation and parts-per-quadrillion determination of fullerenes in river samples from Barcelona catchment area. *Anal. Bioanal. Chem.* **2015**, *407*, 4261–4275. [CrossRef]

9. Sanchís, J.; Milačič, R.; Zuliani, T.; Vidmar, J.; Abad, E.; Farré, M.; Barceló, D. Occurrence of C_{60} and related fullerenes in the Sava River under different hydrologic conditions. *Sci. Total Environ.* **2018**, *643*, 1108–1116. [CrossRef]

10. Kovač, T.; Šarkanj, B.; Klapec, T.; Borišev, I.; Kovač, M.; Nevistić, A.; Strelec, I. Fullerol $C_{60}(OH)_{24}$ nanoparticles and mycotoxigenic fungi: A preliminary investigation into modulation of mycotoxin production. *Environ. Sci. Pollut. Res.* **2017**, *24*, 16673–16681. [CrossRef]

11. Kovač, T.; Šarkanj, B.; Klapec, T.; Borišev, I.; Kovač, M.; Nevistić, A.; Strelec, I. Antiaflatoxigenic effect of fullerene C60nanoparticles at environmentally plausible concentrations. *AMB Express* **2018**, *8*. [CrossRef] [PubMed]

12. Kovač, T.; Borišev, I.; Crevar, B.; Čačić Kenjerić, F.; Kovač, M.; Strelec, I.; Ezekiel, C.N.; Sulyok, M.; Krska, R.; Šarkanj, B.; et al. Fullerol $C_{60}(OH)_{24}$ nanoparticles modulate aflatoxin B1 biosynthesis in Aspergillus flavus. *Sci. Rep.* **2018**, *8*, 12855. [CrossRef] [PubMed]

13. Kovač, T.; Borišev, I.; Kovač, M.; Lončarić, A.; Čačić Kenjerić, F.; Djordjevic, A.; Strelec, I.; Ezekiel, C.N.; Sulyok, M.; Krska, R.; et al. Impact of fullerol $C_{60}(OH)_{24}$ nanoparticles on the production of emerging toxins by Aspergillus flavus. *Sci. Rep.* **2020**, *10*. [CrossRef] [PubMed]

14. Pinotti, L.; Ottoboni, M.; Giromini, C.; Dell'Orto, V.; Cheli, F. Mycotoxin contamination in the EU feed supply chain: A focus on Cereal Byproducts. *Toxins* **2016**, *8*, 45. [CrossRef] [PubMed]

15. Eskola, M.; Kos, G.; Elliott, C.T.; Hajšlová, J.; Mayar, S.; Krska, R. Worldwide contamination of food-crops with mycotoxins: Validity of the widely cited 'FAO estimate' of 25%. *Crit. Rev. Food Sci. Nutr.* **2019**. [CrossRef] [PubMed]

16. Kovač, M.; Šubarić, D.; Bulaić, M.; Kovač, T.; Šarkanj, B. Yesterday masked, today modified; what do mycotoxins bring next? *Arh. Hig. Rada Toksikol.* **2018**, *69*, 196–214. [CrossRef] [PubMed]

17. EU Comission. *RASFF—The Rapid Alert System for Food and Feed—2018 Annual Report More*; EU Comission: Brussels, Belgium, 2018.

18. European Commission. *Commission Regulation (EC) No 1881/2006 Setting Maximum Levels for Certain Contaminants in Foodstuffs*; European Commission: Brussels, Belgium, 2006.

19. The European Commission. Recomendations on the presence of T-2 and HT-2 toxin in cereals and cereal products. *Off. J. Eur. Union* **2013**, *56*, 12–15. [CrossRef]

20. EFSA. Deoxynivalenol in food and feed: Occurrence and exposure. *EFSA J.* **2013**, *11*, 3379–3434. [CrossRef]

21. EFSA. Appropriateness to set a group health-based guidance value for zearalenone and its modified forms. *EFSA J.* **2016**, *14*, 4425. [CrossRef]

22. Medina, A.; Akbar, A.; Baazeem, A.; Rodriguez, A.; Magan, N. Climate change, food security and mycotoxins: Do we know enough? *Fungal Biol. Rev.* **2017**, *31*, 143–154. [CrossRef]

23. Bebber, D.P.; Ramotowski, M.A.T.; Gurr, S.J. Crop pests and pathogens move polewards in a warming world. *Nat. Clim. Chang.* **2013**, *3*, 985–988. [CrossRef]

24. Helfer, S. Rust fungi and global change. *New Phytol.* **2014**, *201*, 770–780. [CrossRef] [PubMed]

25. Trnka, M.; Rötter, R.P.; Ruiz-Ramos, M.; Kersebaum, K.C.; Olesen, J.E.; Žalud, Z.; Semenov, M.A. Adverse weather conditions for European wheat production will become more frequent with climate change. *Nat. Clim. Chang.* **2014**, *4*, 637–643. [CrossRef]

26. Medina, Á.; Rodríguez, A.; Magan, N. Climate change and mycotoxigenic fungi: Impacts on mycotoxin production. *Curr. Opin. Food Sci.* **2015**, *5*, 99–104. [CrossRef]

27. Battilani, P.; Toscano, P.; Van Der Fels-Klerx, H.J.; Moretti, A.; Leggieri, M.C.; Brera, C. Aflatoxin B 1 contamination in maize in Europe increases due to climate change. *Sci. Rep.* **2016**, *6*, 24328. [CrossRef]

28. Ponts, N. Mycotoxins are a component of Fusarium graminearum stress-response system. *Front. Microbiol.* **2015**, *6*, 1234. [CrossRef]

29. Narasaiah, K.V.; Sashidhar, R.B.; Subramanyam, C. Biochemical analysis of oxidative stress in the production of aflatoxin and its precursor intermediates. *Mycopathologia* **2006**, *162*, 179–189. [CrossRef]

30. Sachkova, A.S.; Kovel, E.S.; Churilov, G.N.; Guseynov, O.A.; Bondar, A.A.; Dubinina, I.A.; Kudryasheva, N.S. On mechanism of antioxidant effect of fullerenols. *Biochem. Biophys. Rep.* **2017**, *9*, 1–8. [CrossRef]

31. Jović, D.S.; Seke, M.N.; Djordjevic, A.N.; Mrdanović, J.; Aleksić, L.D.; Bogdanović, G.M.; Pavić, A.B.; Plavec, J. Fullerenol nanoparticles as a new delivery system for doxorubicin. *RSC Adv.* **2016**, *6*, 38563–38578. [CrossRef]

32. Vraneš, M.; Borišev, I.; Tot, A.; Armaković, S.; Armaković, S.; Jović, D.; Gadžurić, S.; Djordjevic, A. Self-assembling, reactivity and molecular dynamics of fullerenol nanoparticles. *Phys. Chem. Chem. Phys.* **2017**, *19*, 135–144. [CrossRef]

33. Borišev, M.; Borišev, I.; Župunski, M.; Arsenov, D.; Pajević, S.; Ćurćić, Ž.; Vasin, J.; Djordjevic, A. Drought impact is alleviated in sugar beets (*Beta vulgaris* L.) by foliar application of fullerenol nanoparticles. *PLoS ONE* **2016**, *11*, e0166248. [CrossRef] [PubMed]

34. Rauscher, H.; Roebben, G.; Rauscher, H.; Roebben, G.; Sanfeliu, A.B.; Emons, H.; Gibson, N.; Koeber, R.; Linsinger, T.; Rasmussen, K.; et al. *Towards a Review of the EC Recommendation for a Definition of the Term "Nanomaterial", Part 3: Scientific-Technical Evaluation of Options to Clarify the Definition and to Facilitate Its Implementation*; EU Commision: Brussels, Belgium, 2015.

35. Holmes, R.A.; Boston, R.S.; Payne, G.A. Diverse inhibitors of aflatoxin biosynthesis. *Appl. Microbiol. Biotechnol.* **2008**, *78*, 559–572. [CrossRef] [PubMed]

36. Medina, A.; Rodriguez, A.; Magan, N. Effect of climate change on *Aspergillus flavus* and aflatoxin B1 production. *Front. Microbiol.* **2014**, *5*, 348. [CrossRef] [PubMed]

37. Fontaine, K.; Mounier, J.; Coton, E.; Hymery, N. Individual and combined effects of roquefortine C and mycophenolic acid on human monocytic and intestinal cells. *World Mycotoxin J.* **2016**, *9*, 51–61. [CrossRef]

38. JH, D. The occurrence, properties and significance of citrinin mycotoxin. *J. Plant Pathol. Microbiol.* **2015**, *6*. [CrossRef]

39. European Food Safety Authority (EFSA). Scientific opinion on the risks for public and animal health related to the presence of citrinin in food and feed. *EFSA J.* **2012**, *10*, 2605. [CrossRef]

40. Šarkanj, B.; Ezekiel, C.N.; Turner, P.C.; Abia, W.A.; Rychlik, M.; Krska, R.; Sulyok, M.; Warth, B. Ultra-sensitive, stable isotope assisted quantification of multiple urinary mycotoxin exposure biomarkers. *Anal. Chim. Acta* **2018**, *1019*, 84–92. [CrossRef] [PubMed]

41. Gotthardt, M.; Asam, S.; Gunkel, K.; Moghaddam, A.F.; Baumann, E.; Kietz, R.; Rychlik, M. Quantitation of six Alternaria toxins in infant foods applying stable isotope labeled standards. *Front. Microbiol.* **2019**, *10*. [CrossRef]

42. Lind, A.L.; Wisecaver, J.H.; Smith, T.D.; Feng, X.; Calvo, A.M.; Rokas, A. Examining the evolution of the regulatory circuit controlling secondary metabolism and development in the fungal genus aspergillus. *PLoS Genet.* **2015**, *11*, 1005096. [CrossRef]

43. Xu, Y.; Vinas, M.; Alsarrag, A.; Su, L.; Pfohl, K.; Rohlfs, M.; Schäfer, W.; Chen, W.; Karlovsky, P. Bis-naphthopyrone pigments protect filamentous ascomycetes from a wide range of predators. *Nat. Commun.* **2019**, *10*, 1–12. [CrossRef]

44. Abass, A.B.; Awoyale, W.; Sulyok, M.; Alamu, E.O. Occurrence of regulated mycotoxins and other microbial metabolites in dried cassava products from nigeria. *Toxins* **2017**, *9*, 207. [CrossRef] [PubMed]

45. Wauters, I.; Goossens, H.; Delbeke, E.; Muylaert, K.; Roman, B.I.; Van Hecke, K.; Van Speybroeck, V.; Stevens, C.V. Beyond the diketopiperazine family with alternatively bridged Brevianamide F analogues. *J. Org. Chem.* **2015**, *80*, 8046–8054. [CrossRef] [PubMed]

46. Weber, J.; Vaclavikova, M.; Wiesenberger, G.; Haider, M.; Hametner, C.; Fröhlich, J.; Berthiller, F.; Adam, G.; Mikula, H.; Fruhmann, P.; et al. Chemical synthesis of culmorin metabolites and their biologic role in culmorin and acetyl-culmorin treated wheat cells. *Org. Biomol. Chem.* **2018**, *16*, 2043–2048. [CrossRef] [PubMed]

47. Woelflingseder, L.; Warth, B.; Vierheilig, I.; Schwartz-Zimmermann, H.; Hametner, C.; Nagl, V.; Novak, B.; Šarkanj, B.; Berthiller, F.; Adam, G.; et al. The Fusarium metabolite culmorin suppresses the in vitro glucuronidation of deoxynivalenol. *Arch. Toxicol.* **2019**, *93*, 1729–1743. [CrossRef] [PubMed]

48. Burke, L.T.; Dixon, D.J.; Ley, S.V.; Rodríguez, F. Total synthesis of the Fusarium toxin equisetin. *Org. Biomol. Chem.* **2005**, *3*, 274–280. [CrossRef]

49. Beccari, G.; Colasante, V.; Tini, F.; Senatore, M.T.; Prodi, A.; Sulyok, M.; Covarelli, L. Causal agents of Fusarium head blight of durum wheat (*Triticum durum* Desf.) in central Italy and their in vitro biosynthesis of secondary metabolites. *Food Microbiol.* **2018**, *70*, 17–27. [CrossRef] [PubMed]

50. Sondergaard, T.E.; Hansen, F.T.; Purup, S.; Nielsen, A.K.; Bonefeld-Jørgensen, E.C.; Giese, H.; Sørensen, J.L. Fusarin C acts like an estrogenic agonist and stimulates breast cancer cells in vitro. *Toxicol. Lett.* **2011**, *205*, 116–121. [CrossRef]

51. Dhani, S.; Nagiah, S.; Naidoo, D.B.; Chuturgoon, A.A. Fusaric Acid immunotoxicity and MAPK activation in normal peripheral blood mononuclear cells and Thp-1 cells. *Sci. Rep.* **2017**, *7*, 3051. [CrossRef]

52. Beccari, G.; Arellano, C.; Covarelli, L.; Tini, F.; Sulyok, M.; Cowger, C. Effect of wheat infection timing on Fusarium head blight causal agents and secondary metabolites in grain. *Int. J. Food Microbiol.* **2019**, *290*, 214–225. [CrossRef]

53. Hennig-Pauka, I.; Koch, F.J.; Schaumberger, S.; Woechtl, B.; Novak, J.; Sulyok, M.; Nagl, V. Current challenges in the diagnosis of zearalenone toxicosis as illustrated by a field case of hyperestrogenism in suckling piglets. *Porc. Health Manag.* **2018**, *4*, 18. [CrossRef]

54. Liuzzi, V.C.; Mirabelli, V.; Cimmarusti, M.T.; Haidukowski, M.; Leslie, J.F.; Logrieco, A.F.; Caliandro, R.; Fanelli, F.; Mulè, G. Enniatin and beauvericin biosynthesis in Fusarium species: Production profiles and structural determinant prediction. *Toxins* **2017**, *9*, 45. [CrossRef] [PubMed]

55. Malachová, A.; Sulyok, M.; Beltrán, E.; Berthiller, F.; Krska, R. Optimization and validation of a quantitative liquid chromatography-tandem mass spectrometric method covering 295 bacterial and fungal metabolites including all regulated mycotoxins in four model food matrices. *J. Chromatogr. A* **2014**, *1362*, 145–156. [CrossRef] [PubMed]

56. Djordjević, A. Fullerene science and technology catalytic preparation and characterization of $C_{60}Br_{24}$. *Fuller. Sci. Technol.* **1998**, *6*, 689–694. [CrossRef]

57. Mirkov, S.M.; Djordjevic, A.N.; Andric, N.L.; Andric, S.A.; Kostic, T.S.; Bogdanovic, G.M.; Vojinovic-Miloradov, M.B.; Kovacevic, R.Z. Nitric oxide-scavenging activity of polyhydroxylated fullerenol, $C_{60}(OH)_{24}$. *Nitric Oxide Biol. Chem.* **2004**, *11*, 201–207. [CrossRef]

58. Šarkanj, B.; Molnar, M.; Čačić, M.; Gille, L. 4-Methyl-7-hydroxycoumarin antifungal and antioxidant activity enhancement by substitution with thiosemicarbazide and thiazolidinone moieties. *Food Chem.* **2013**, *139*, 488–495. [CrossRef]

59. Jerković, I.; Kranjac, M.; Marijanović, Z.; Šarkanj, B.; Cikoš, A.M.; Aladić, K.; Pedisić, S.; Jokić, S. Chemical Diversity of Codium bursa (Olivi) C. Agardh headspace compounds, volatiles, fatty acids and insight into its antifungal activity. *Molecules* **2019**, *24*, 842. [CrossRef]

MDPI
St. Alban-Anlage 66
4052 Basel
Switzerland
Tel. +41 61 683 77 34
Fax +41 61 302 89 18
www.mdpi.com

Toxins Editorial Office
E-mail: toxins@mdpi.com
www.mdpi.com/journal/toxins

www.ingramcontent.com/pod-product-compliance
Lightning Source LLC
LaVergne TN
LVHW070642170726
843515LV00004B/306